CliffsQuickReview®
Math Word Problems

By Karen L. Anglin

Course Hero

CliffsQuickReview® Math Word Problems

2nd Edition

By Karen L. Anglin

About the Author

Karen Anglin has been a mathematics instructor at Blinn College in Brenham, Texas, since 1990. She has given numerous award-winning presentations to educators on the topic of math word problems. Mrs. Anglin has a B.S. degree in Mathematics with a secondary teaching certificate.

Publisher's Acknowledgments

Editorial

Senior Acquisitions Editor: Greg Tubach
Project Editor: Marcia L. Johnson
Copy Editor: Tere Drenth
Technical Editor: David Herzog
Editorial Assistant: Amanda Harbin

Production

Indexer: Joan Griffitts
Proofreader: Linda Quigley
Wiley Publishing, Inc. Composition Services

CliffsQuickReview® Math Word Problems
Published by:
Course Hero, Inc.
2000 Seaport Blvd, 3rd Floor
Redwood City, CA 94063
United States of America

Published by Course Hero

ISBN: 978-1-957671-05-5

Printed in the United States of America

10 9 8 7 6 5 4 3 2 1

1O/SY/QU/QU/IN

Table of Contents

INTRODUCTION

CliffsQuickReview *Math Word Problems* is designed to give a clear, concise, easy-to-use review of the basics of solving math word problems. Introducing each topic, defining key terms, and carefully "walking" through each sample problem type in a step-by-step manner gives you insight and understanding to solving math word problems. The goal of this book is to guide you through successful basic algebraic problem solving.

The prerequisite to benefit the most from this book is some basic pre-algebra skills and knowledge—knowing what a variable is, being able to use the order of operations, being able to perform arithmetic with negative numbers, being able to find the least common multiple, and being familiar with solving a simple equation. The process of solving a simple equation is reviewed in detail. If you need to be reminded of some of the terminology and properties, use the Glossary at the back of this book.

One of the skills you develop by reading this book is the ability to identify what type of word problems you are asked to solve. Another skill is the ability to match a word problem with the appropriate problem solving strategy. For this reason, at the end of each chapter, you are given review problems of several different types. These can be challenging and will test what you have learned. Remember that, in life, problems are not in section order!

Chapter 4 is the first chapter in which you begin solving story problems. The first three chapters set a firm foundation upon which you build your skills for solving word problems.

Why You Need This Book

Can you answer yes to any of these questions?

- Do you need to review math word problems?
- Do you need a course supplement to a math class?
- Do you need a concise, comprehensive reference for math word problems?
- Do you want to have more confidence when faced with a word problem?

If so, then CliffsQuickReview *Math Word Problems* is for you!

How to Use This Book

You can use this book in any way that fits your personal style for study and review—you decide what works best with your needs. You can either read the book from cover to cover or just look for the information you want and put it back on the shelf for later. Here are a few ways you can search for topics:

- Use the Pocket Guide to find essential information, such as keywords to identify operations, keywords for translating in the correct order, indications to insert parentheses, ways to solve equations, and ways to choose the appropriate strategy for the word problem.
- Look for areas of interest in the book's Table of Contents, or use the index to find specific topics.
- Flip through the book looking for subject areas at the top of each page.
- Get a glimpse of what you'll gain from a chapter by reading through the "Chapter Check-In" at the beginning of each chapter.
- Use the "Chapter Checkout" at the end of each chapter to gauge your grasp of the important information you need to know.
- Test your knowledge more completely in the CQR Review and look for additional sources of information in the CQR Resource Center.
- Use the Glossary to find key terms fast. This book defines new terms and concepts where they first appear in the chapter. If a word is **boldfaced,** you can find a more complete definition in the book's Glossary.
- Or flip through the book until you find what you're looking for—we organized this book to gradually build on key concepts.

Visit Our Web Site

A great resource, `www.cliffsnotes.com`, features timely articles and tips, plus downloadable versions of many CliffsNotes books.

When you stop by our site, don't hesitate to share your thoughts about this book or any other Wiley product. Just click the Talk to Us button. We welcome your feedback!

Chapter 1

TRANSLATING EXPRESSIONS

Chapter Check-In

- ❑ Addition and subtraction keywords
- ❑ Multiplication and division keywords
- ❑ Turnaround words

The first step in solving a word problem is always to read the problem. You need to be able to **translate** words into mathematical symbols, focusing on **keywords** that indicate the mathematical procedures required to solve the problem—both the operation and the order of the expression. In much the same way that you can translate Spanish into English, you can translate English words into symbols, the language of mathematics. Many (if not all) keywords that indicate mathematical operations are familiar words.

Keywords of Basic Mathematical Operations

To begin, you translate English phrases into algebraic **expressions.** An algebraic expression is a collection of numbers, **variables,** operations, and **grouping symbols.** You will translate an unknown number as the variable *x* or *n.* The grouping symbols are usually a set of parentheses, but they can also be sets of brackets or braces.

In translating expressions, you want to be well acquainted with basic keywords that translate into mathematical operations: addition keywords, subtraction keywords, multiplication keywords, and division keywords, which are covered in the four following sections.

Addition keywords

Some common examples of addition keywords are as follows:

SUM OF______ AND ______
TOTAL OF ______ AND ______
______ PLUS ______
______ INCREASED BY ______
GAIN
RAISE
MORE
INCREASE OF

The first two keywords (SUM and TOTAL) are called **leading keywords** because they lead the expression. The second two keywords (PLUS and INCREASED BY) are keywords that indicate the exact placement of the plus sign. The last four keywords can be found in word problems and may indicate addition.

When an expression begins with the leading keywords SUM or TOTAL, the leading keyword defines the corresponding AND. The plus sign then physically replaces the AND in the expression.

Example 1: Translate the following: the sum of five and a number

The following steps help you translate this problem:

1. **Underline the words before and after AND when it corresponds to the leading keyword SUM OF.**

   the sum of five and a number

2. **Circle the leading keyword and indicate the corresponding AND that it defines.**

   the (sum of) five and a number

3. **Translate each underlined expression and replace AND with a plus sign.**

   The expression translates to $5 + x$.

Example 2: Translate the following: the total of a number and negative three

Use the following steps to translate this problem:

1. **The keyword TOTAL OF is a leading keyword that defines AND, so underline the words before and after AND: "a number" and "negative three."**

   the total of a number and negative three

2. **Circle the leading keyword and indicate the corresponding AND that it defines.**

   the (total of) a number and negative three

3. **Translate each underlined expression and replace AND with a plus sign.**

   The expression translates to $x + -3$.

Example 3: Translate the following: the sum of seven and negative four

Translate this example in the following way:

1. **The word SUM OF is a leading keyword that defines AND , so underline the words before and after AND: "seven" and "negative four."**

   the sum of seven and negative four

2. **Circle the leading keyword and indicate the corresponding AND that it defines.**

   the (sum of) seven and negative four

3. **Translate each underlined expression and replace AND with a plus sign.**

   The expression translates to $7 + -4$.

Reminder: The AND keyword translates to mean "plus" because the leading keyword is SUM OF. With other leading keywords (discussed in the following sections), AND can mean other things. Also notice that you do not simplify the expression and get "3" for the answer because you are just translating words into symbols and not performing the math.

Two other keywords on the addition keyword list, PLUS and INCREASED BY, can be correctly translated by the **direct translation strategy.** In the direct translation strategy, you translate each word into its corresponding algebraic symbol, one at a time, in the same order as written, as shown in Example 4.

Example 4: Translate the following: a number increased by twenty-four

The expression translates to $x + 24$.

Some additional keywords, such as GAIN, MORE, INCREASE OF, and RAISE, are commonly found in story problems, as in Example 5.

Example 5: Translate the following story problem into a mathematical expression about the weight of the linebacker: The defensive linebacker weighed two hundred twenty-two pounds at the beginning of spring training. He had a gain of seventeen pounds after working out with the team for four weeks.

The expression translates to $222 + 17$.

Note: Not all numbers mentioned in a word problem should be included in the mathematical expression. The number "four" is just interesting fact, but it is not information you need in order to write an expression about the linebacker's weight.

You may also be wondering why the answer isn't 239 pounds. That's because the question asks you to translate the story problem into a mathematical expression, not to evaluate the expression.

Example 6: Translate the following word problem into a mathematical expression about the cashier's current hourly wage: A cashier at the corner grocery was earning $6.25 an hour. He received a raise of 25 cents an hour.

The expression translates to $6.25 + 0.25$.

Note: The hourly wage is stated in dollars, and the raise is stated in cents. Any time you are adding two numbers that have **units,** make sure both numbers are measured with the same units; if they aren't, convert one of the numbers to the same units as the other. Having both numbers measured with the same units is called **homogeneous units.** In this example, you convert his raise, the 25 cents, to $0.25 because his hourly wage is measured in dollars, not cents, so the raise must also be in dollars.

Subtraction keywords

Subtraction keywords also include leading keywords, keywords that can be translated one word at a time, and keywords that are found in story problems. Look at the following list of subtraction keywords:

DIFFERENCE BETWEEN ______ AND ______

______ MINUS ______

______ DECREASED BY ______

LOSS

LESS

FEWER

TAKE AWAY

One subtraction keyword (DIFFERENCE BETWEEN) is a two-part expression that begins with a leading keyword that defines the corresponding AND. You can use the same methods of underlining and circling the keywords shown in the preceding section to translate these expressions.

Example 7: Translate the following: the difference between four and six

Here is how you translate Example 7:

1. **Because the keyword DIFFERENCE BETWEEN is a leading keyword that defines the corresponding AND, underline the words before and after AND: "four" and "six."**

   the difference between four and six

2. **Circle the leading keyword and indicate the corresponding AND that it defines.**

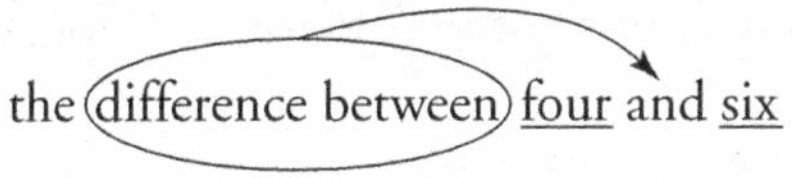

3. **Translate each underlined expression and replace AND with a minus sign.**

   The expression translates to $4 - 6$.

Note: AND is not always translated to mean addition. Here, the DIFFERENCE BETWEEN is the leading keyword that defines the AND to mean subtraction.

Other subtraction keywords, such as MINUS and DECREASED BY, use the direct translation strategy. Example 8 is a subtraction word problem that is translated one keyword at a time, in the exact order of the expression.

Example 8: Translate the following: twenty-four decreased by a number

The expression translates to $24 - x$.

In a story problem, you may find the subtraction keywords LOSS, LESS, FEWER, and TAKE AWAY, as shown in Example 9.

Example 9: Translate the following word problem into a mathematical expression about the current value of materials at the job site: A construction company stored $1,253 worth of materials at the job site. The company suffered a loss of $300 due to storm damage.

The expression translates to 1,253 – 300.

Multiplication keywords

Some common examples of multiplication keywords are as follows:

MULTIPLY ______ BY ______

PRODUCT OF ______ AND ______

______ TIMES ______

DOUBLE ______

TWICE ______

TRIPLE ______

PERCENT OF ______

FRACTION OF ______

For two of the multiplication keywords, MULTIPLY and PRODUCT OF, a leading keyword defines the corresponding BY or AND, as shown in Example 10.

Example 10: Translate the following: the product of seven and a number

Translate this example in the following way:

1. **Because PRODUCT OF is a leading keyword that corresponds to AND, underline the words before and after AND: "seven" and "a number."**

   the product of seven and a number

2. **Circle the leading keyword and indicate the corresponding AND that it defines.**

   the (product of) seven and a number

3. **Translate each underlined expression and replace AND with a times sign.**

   The expression translates to $7 \times x$.

Note: Keep in mind that AND does not always indicate addition. The keyword PRODUCT OF defines the AND in this expression to mean multiplication.

A multiplication expression that is translated by the direct translation method is shown in Example 11.

Example 11: Translate the following: a number times fifteen

The expression translates to $x \times 15$.

Some multiplication keywords, such as DOUBLE, TWICE, and TRIPLE, translate into a number and the operation of multiplication, as shown in Examples 12 and 13.

Example 12: Translate the following: twice a number

The expression translates to $2 \times x$.

Example 13: Translate the following word problem into a mathematical expression: Jennifer had \$15 dollars in the bank. Over the next two weeks she doubled her money.

The expression translates to 2×15.

One of the keywords that indicates multiplication is OF. In word problems, however, you may see more than one use of the word "of." The only OF that indicates multiplication is the one that follows the keyword PERCENT, the percent sign, the keyword FRACTION, or a fraction. See Examples 14 and 15.

Example 14: Translate the following: twenty five percent of four hundred dollars

The expression translates to 0.25×400.

Note: Remember that a percent is changed to a decimal before multiplying.

Example 15: Translate the following: one-third of twenty-seven

The expression translates to $\frac{1}{3} \times 27$.

Division keywords

Some common examples of division keywords are as follows:

QUOTIENT OF ______ AND ______
DIVIDE ______ BY ______

______ DIVIDED BY ______
DIVIDED EQUALLY
PER

The keywords PRODUCT OF and QUOTIENT OF are difficult for some people to differentiate. Here is a hint to help you remember which one indicates division and which one indicates multiplication: Quotient is a "harder" word than "product," and division is a "harder" operation than multiplication.

Remember: Leading keywords define the corresponding AND or BY to mean divide, usually designated with the symbol ÷.

Example 16: Translate the following: the quotient of seven and a number

1. **Because the keyword QUOTIENT OF is a leading keyword that defines AND, underline the words before and after AND: "seven" and "a number."**

   the quotient of seven and a number

2. **Circle the leading keyword and indicate the corresponding AND that it defines.**

   the quotient of seven and a number

3. **Translate each underlined expression and replace AND with a division sign.**

   The expression translates to $7 \div n$.

Note: Here, the keyword QUOTIENT OF defines AND to mean division.

Example 17: Translate the following: divide negative thirty-six by nine

1. **Because the word DIVIDE is a leading keyword that defines the BY, underline the words before and after BY: "negative thirty-six" and "nine."**

   divide negative thirty-six by nine

2. **Circle the leading keyword and indicate the corresponding BY that it defines.**

   divide negative thirty-six by nine

3. **Translate each underlined expression and replace AND with a division sign.**

   The expression translates to $-36 \div 9$, $\frac{-36}{9}$, or $9\overline{)-36}$.

Note: The first number goes in the numerator when using a fraction bar to indicate division. The number in the numerator (the –36) goes inside the "house" when using the long division symbol.

Some division keywords can be translated one word at a time. Instead, you just follow the sentence and replace with algebraic notations as you go along.

Example 18: Translate the following: a number divided by 16

The expression translates to $x \div 16$ or $\frac{x}{16}$.

Often, in story problems, the keyword that indicates division is PER. When a story problem asks for the speed of a vehicle in miles per hour, set up the expression to divide the number of miles by the number of hours. You not only directly translate "miles" ÷ "hours," but also identify the number of miles and number of hours by finding them elsewhere in the problem. See Example 19.

Example 19: Translate the following word problem into a mathematical expression about speed: It takes three hours to travel 150 miles to grandmother's house. How do you find your average speed in miles per hour?

You find "miles" ÷ "hours" in the question. In the first part of the word problem, you find the number of miles, 150 miles, and the number of hours, three hours.

The expression translates to $150 \div 3$.

Keywords That Indicate a Change in Order

Some keywords are not included in other lists in this chapter because they are a bit different from other types of keywords. This section gives you additional keywords, called **turnaround words,** and these indicate a change in order from the original English phrase.

Throughout this chapter, you use keywords in one of two ways:

- Translate the words directly, in the order they are given.
- Recognize leading keywords and find the corresponding AND, TO, BY, or FROM that tell you how to translate the equation.

To help you understand turnaround words, think about the directions on a box of cake mix. If the directions say "three eggs added to the mix," which do you put into the bowl first? You put the mix into the bowl first, and then add the eggs. The word TO is one of the basic turnaround words discussed in the following section, and to help you remember to turn the expression around, you box the word.

three eggs added [to] the mix

Basic turnaround words

Certain keywords indicate a turn around in the order of the translation. All of the keywords indicating a change in order contain the following words:

TO
FROM
THAN

Addition turnaround words

Addition keywords that indicate a turnaround are

ADD ______ TO ______
______ ADDED TO ______
______ MORE THAN ______

Example 20: Translate the following: twelve added to negative four

To help you translate this problem, box the turnaround word.

twelve added [to] negative four

Replace each word with algebraic symbols and turn the expression around.

The expression translates to $-4 + 12$.

Example 21: Translate the following: add negative three to five

Example 21 uses both a turnaround word and a leading keyword, and it is translated as follows:

1. **Because the first word in the expression, ADD, indicates an operation, ADD is a leading keyword. ADD defines the TO, so underline the words before and after TO: "negative three" and "five."**

   add <u>negative three</u> to <u>five</u>

2. **Circle the leading keyword and indicate the corresponding TO that it defines; box the turnaround word, TO.**

add negative three to five

3. **Translate each underlined expression, replace T[illegible] sign, and turn the expression around.**

   The expression translates to 5 + –3.

People sometimes argue that a turnaround word is not necessary with addition because of the **Commutative Property of Addition;** that is, both –3 + 5 and 5 + –3 result in the same answer for Example 21 (+2) when **simplified.** However, the latter (5 + –3) shows an understanding of the order indicated by the keyword. You have two reasons to learn to translate expressions in the correct order:

- Good habits are formed for correct translation of subtraction and division expressions, which are not commutative.
- Just as the makers of a cake mix intend for you to put the cake mix in the bowl first and add the eggs later, the author of the problem intends to have you perform the addition in the prescribed order.

Subtraction turnaround words

Subtraction keywords that indicate a turnaround are

SUBTRACT ______ FROM ______

______ SUBTRACTED FROM ______

______ LESS THAN ______

Example 22: Translate the following: a number less than seven

To help you solve this problem, box the turnaround word, THAN.

a number less than seven

Replace each word with algebraic symbols and turn the expression around.

The expression translates to 7 – *x*.

Example 23: Translate the following: subtract seventeen from fifty-four

1. **Because the word SUBTRACT is a leading keyword that defines FROM, underline the words before and after FROM: "seventeen" and "fifty-four."**

   subtract seventeen from fifty-four

Circle the leading keyword and indicate the corresponding FROM that it defines; box the turnaround word, FROM.

(subtract) seventeen [from] fifty-four

3. **Translate each underlined expression, replace FROM with a minus sign, and turn the expression around.**

   The expression translates to 54 – 17.

Multiplication turnaround words

None of the multiplication keywords indicates a turnaround. All multiplication expressions can be translated using the direct translation strategy or leading keywords.

The product of a number and 8 can be translated $n \times 8$, but most often, you see the expression written as $8 \times n$ or $8n$, because mathematicians have set a standard that the **coefficient** is written before the variable. (In this example, the number 8 is the coefficient.)

Note: The expression $8n$ uses **implied multiplication.** Multiplication is implied when a number is placed next to a variable, or when a number is placed next to an expression surrounded by parentheses. Although a multiplication sign is not shown, its use is implied.

Division turnaround words

Division keywords that indicate a turnaround are:

DIVIDE ______ INTO ______

______ DIVIDED INTO ______

Notice that TO, a basic turnaround word, is included in the word INTO and indicates a turnaround.

Example 24: Translate the following: divide five into 125

Solve Example 24 as follows:

1. **Because the word DIVIDE is a leading keyword that defines INTO, underline the words before and after INTO: "five" and "125."**

   divide five into 125

2. **Circle the leading keyword and indicate the corresponding INTO that it defines; box the turnaround word, INTO.**

divide five into 125

3. **Translate each underlined expression, replace INTO with a division sign, and turn the expression around.**

   The expression translates to 125 ÷ 5.

Example 25: Translate the following: twenty-five divided into one hundred

To help you translate this problem, box the turnaround word, INTO.

twenty-five divided into 100

Replace each word with algebraic symbols and turn the expression around.

The expression translates to 100 ÷ 25.

Chapter Checkout

Q&A

Translate the following English phrases into algebraic expressions or fill in the blanks.

1. The sum of six and a number
2. Seven plus five
3. The difference between a number and eleven
4. Three minus a number
5. Multiply four by a number
6. Ten percent of twenty
7. One fifth of forty-five
8. Triple a number
9. A number divided by eight
10. The quotient of forty-nine and a number
11. The basic turnaround words are ______, ______, and ______.
12. Add four to a number
13. A number added to three
14. Seven more than five

15. Subtract three from four

16. Thirty-six less than a number

17. Divide five into twenty-five

Answers: **1.** $6 + x$ **2.** $7 + 5$ **3.** $x - 11$ **4.** $3 - x$ **5.** $4x$ **6.** 0.10×20 **7.** $\frac{1}{5} \times 45$ **8.** $3x$ **9.** $x \div 8$ **10.** $49 \div x$ **11.** TO, FROM, THAN **12.** $x + 4$ **13.** $3 + x$ **14.** $5 + 7$ **15.** $4 - 3$ **16.** $x - 36$ **17.** $25 \div 5$

Chapter 2
INSERTING PARENTHESES

Chapter Check-In

- ❑ Translating a comma
- ❑ Leading keywords
- ❑ Adjacent keywords
- ❑ Turnaround words

Expressions with **multiple operations** may need parentheses in order to be properly translated. You have four ways to tell when parentheses are needed, and they are discussed throughout this chapter.

First, however, look at the three following examples, which review the techniques for translating expressions with multiple operations that do not require parentheses.

Example 1: Translate the following: a number plus seven minus four

The expression translates to $x + 7 - 4$.

Example 2: Translate the following: twenty-one decreased by three times a number

The expression translates to $21 - 3x$.

Example 3: Translate the following: twice a number divided by nine

The expression translates to $2x \div 9$.

Translating a Comma and IS

A comma in an expression indicates the completion of one operation and the beginning of another. The word IS also acts as a separator in the same way that a comma does. Put parentheses around the expression that precedes the comma or IS.

Example 4: Translate the following: the sum of thirteen and a number, subtracted from four

Translate Example 4 as follows:

1. **Because a comma follows the word "number," put parentheses around the expression that ends with the word "number."**

   (the sum of thirteen and a number), subtracted from four

   Do you notice the leading keyword, SUM OF, that defines the corresponding AND? You underline the words before and after AND.

2. **Underline "thirteen" and "a number."**

   (the sum of thirteen and a number), subtracted from four

3. **Circle the leading keyword and indicate the AND that it defines.**

   Do you notice the turnaround word, FROM?

   (the sum of thirteen and a number), subtracted from four

4. **Box the turnaround word to remind you to turn the expression around.**

   (the sum of thirteen and a number), subtracted from four

5. **Translate each underlined expression, replace AND with a plus sign, and turn around the expression.**

   The expression translates to $4 - (13 + x)$.

Example 5: Translate the following: the difference between a number and five is added to twelve

Here is how you translate Example 5:

1. **Because of the separator IS after the word "five," put parentheses around the expression that ends with the word "five."**

   (the difference between a number and five) is added to twelve

   Remember: The keyword DIFFERENCE BETWEEN is a leading keyword and defines the corresponding AND, so you underline the words before and after AND.

2. **Underline "a number" and "five."**

   (the difference between a number and five) is added to twelve

3. **Circle the leading keyword and indicate the AND that it defines; box the turnaround word.**

   The turnaround word is TO.

   (the difference between a number and five) is added to twelve

4. **Translate each underlined expression, replace AND with a minus sign, and turn around the expression.**

   The expression translates to $12 + (x - 5)$.

In both Examples 4 and 5, notice that a whole phrase is turned around. But not all expressions with parentheses need to be turned around; only those with turnaround words. Example 4 has the word FROM that indicates a change in order. Example 5 has the turnaround word TO. As Chapter 1 tells you, the turnaround words are TO (including INTO), FROM, and THAN.

Example 6: Translate the following: the product of five and a number, minus eight

Translate this problem as follows:

1. **Because of the comma after the word "number," put parentheses around the expression that ends with the word "number."**

   (the product of five and a number), minus eight

2. **Because of the leading keyword PRODUCT OF, underline "five" and "a number," which come before and after AND.**

   (the product of five and a number), minus eight

3. **Circle the leading keyword and indicate the AND that it defines.**

   (the product of five and a number), minus eight

4. **Translate each underlined expression and replace AND with an indication that you are multiplying.**

   The expression translates to $(5n) - 8$.

Are the parentheses necessary in the preceding answer? No. If you follow the order of operations, multiplication will be performed before subtraction, so you can say simply, $5n - 8$. $(5n)$ is an example of placing **harmless parentheses** in the expression; that is, parentheses that are not necessary, but do not do any harm in the expression.

Leading Keywords

When you see a leading keyword, underline the phrases before and after the corresponding AND, TO, BY or FROM, just as in Chapter 1. Then, if the underlined portion also contains a keyword, put parentheses around the underlined expression.

As described in Chapter 1, the following leading keywords indicate addition:

SUM OF ______ AND ______
TOTAL OF ______ AND ______
ADD ______ TO ______

The following leading keywords indicate subtraction:

SUBTRACT ______ FROM ______
DIFFERENCE BETWEEN ______ AND ______

Those that indicate multiplication are as follows:

MULTIPLY ______ BY ______
PRODUCT OF ______ AND ______

And, finally the leading keywords that indicate division are those that follow:

DIVIDE ______ BY ______
DIVIDE ______ INTO ______
QUOTIENT OF ______ AND ______

Leading keywords may indicate the need for parentheses if, after underlining the words around the corresponding AND, FROM, BY, TO and INTO, you see another keyword in the underlined expression. See Example 7.

Example 7: Translate the following: the difference between a number and the sum of a number and four

You translate this problem in the following way:

1. **Underline the expressions before and after the AND that correspond to the leading keyword DIFFERENCE BETWEEN.**

   the difference between a number and the sum of a number and four

2. **Because of the keyword in the second underlined expression, enclose that second expression in parenthesis.**

   the difference between a number and (the sum of a number and four)

3. **Circle the leading keywords and indicate the ANDs that they define.**

   the (difference between) a number and

   (the (sum of) a number and four)

4. **Translate each underlined expression and replace the ANDs with the appropriate subtraction or addition symbol.**

   The expression translates to $x - (x + 4)$.

Example 8: Translate the following: add the difference between four and twice a number to the sum of five and the number

Here is the translation broken down step-by-step:

1. **Because Example 7 contains a leading keyword, ADD, begin by underlining the words before and after the corresponding TO.**

   add the difference between four and twice a number to the sum of five and the number

2. **Because each underlined expression contains a keyword, enclose each expression in parenthesis.**

   add (the difference between four and twice a number) to (the sum of five and the number)

3. **Circle the leading keyword and indicate the TO that it defines; box the turnaround word.**

   (add) (the difference between four and twice

   a number) [to] (the sum of five and the number)

4. Translate each underlined expression individually.

a. In the first underlined expression, underline the terms before and after AND that corresponds to the leading keyword DIFFERENCE BETWEEN.

the difference between four and twice a number

b. Because a keyword, TWICE, appears in the second underlined expression, put parenthesis around that expression.

the difference between four and (twice a number)

c. Circle the leading keyword and indicate the AND that it defines.

the difference between four and (twice a number)

d. Translate each underlined expression and replace AND with a minus sign.

The first underlined expression translates to $[4 - (2x)]$.

Note: Brackets are used when parentheses are nested inside parentheses.

e. Turning to the second underlined expression, follow the same technique: Underline "five" and "a number."

the sum of five and the number

f. Circle the leading keyword and indicate the AND that it defines.

the sum of five and the number

g. Translate each underlined expression and replace AND with a plus sign.

The second underlined expression translates $(5 + x)$.

5. Finally, the expression in the original example also has a turnaround word, TO, so you translate the entire expression to $(5 + x) + [4 - (2x)]$

Example 9: Translate the following: subtract a number plus twelve from the difference between twenty and the number

1. **Underline the expressions before and after FROM, which corresponds to the leading keyword SUBTRACT.**

   subtract a number plus twelve from the difference between twenty and the number

2. **Seeing the keyword in each underlined expression, enclose each expression in parentheses.**

   subtract (a number plus twelve) from (the difference between twenty and the number)

3. **Circle the leading keywords and indicate the FROM and AND that they define; box the turnaround word.**

   subtract (a number plus twelve) from

   (the difference between twenty and the number)

4. **Translate each underlined expression, and then replace FROM with a minus sign. Don't forget to turn the expression around!**

   The expression translates to $(20 - x) - (x + 12)$.

Example 10: Translate the following: the product of seven minus a number and the difference between the same number and four

Translate this word problem in the following way:

1. **Underline the expressions before and after AND, which corresponds with the leading keyword PRODUCT OF.**

   the product of seven minus a number and the difference between the same number and four

2. **Because of the keyword in each underlined expression, enclose each expression in parentheses.**

   the product of (seven minus a number) and (the difference between the same number and four)

3. **Circle the leading keywords and indicate the ANDs that they define.**

   Remember: There is no turnaround in multiplication (see Chapter 1).

the (product of) (seven minus a number) and

(the (difference between) the same number and four)

4. **Translate each underlined expression and replace each AND with the appropriate multiplication or subtraction symbol.**

   The expression translates to $(7 - x) \times (x - 4)$.

Example 11: Translate the following: divide the sum of seven and a number by the same number minus three

Translate this problem as follows:

1. **Underline the expressions before and after BY, which correspond to the leading keyword DIVIDE.**

   divide the sum of seven and a number by the same number minus three

2. **Because each underlined expression contains a keyword, enclose each in parentheses.**

   divide (the sum of seven and a number) by (the same number minus three)

3. **Circle the leading keywords and indicate the BY and AND that they define.**

   (divide) (the (sum of) seven and a number) by

   (the same number minus three)

4. **Translate each underlined expression. Replace AND with a plus sign, and replace BY with a division sign.**

   The expression translates to $(7 + x) \div (x - 3)$.

Adjacent Keywords

Another indicator that you need to insert parentheses in your expression is when you have two **adjacent keywords** (even if a word such as "the" appears between the keywords): You separate the two keywords with an open parenthesis and close the parentheses at the end of the expression.

Example 12: Translate the following: twice the sum of a number and eight

Translate this problem using the following steps.

1. **Notice that TWICE and SUM OF are both keywords indicating operations, multiplication and addition respectively, and are, therefore, adjacent keywords. Place an open parenthesis between the two keywords and a close parenthesis at the end of the expression.**

   twice (the sum of a number and eight)

2. **Underline the expressions before and after AND, which corresponds to the leading keyword SUM OF.**

   twice (the sum of a number and eight)

3. **Circle the leading keyword and indicate the AND that it defines.**

   twice (the sum of a number and eight)

4. **Translate each underlined expression and replace AND with a plus sign.**

   The expression translates to $2(x + 8)$.

Example 13: Translate the following: seven times the difference between twelve and negative four

Translate this problem as follows:

1. **Notice that TIMES and DIFFERENCE BETWEEN both indicate operations, multiplication and subtraction respectively, and are, therefore, adjacent keywords; place an open parenthesis between the two keywords and a close parenthesis at the end.**

   seven times (the difference between twelve and negative four)

2. **Underline the expressions before and after AND, which corresponds to the leading keyword DIFFERENT BETWEEN.**

   seven times (the difference between twelve and negative four)

3. **Circle the leading keyword and indicate the AND that it defines.**

   seven times (the difference between

   twelve and negative four)

4. **Translate each underlined expression and replace AND with a minus sign.**

   The expression translates to $7[12 - (-4)]$. Note that this equation has been translated, but not simplified. In Chapter 3, you learn to simplify expressions.

Turnaround Words

If the keyword includes a turnaround word, such as TO, FROM, or THAN, and the expression(s) that is to be turned around has another keyword, that expression(s) must be enclosed in parentheses before turning it around.

Example 14: Translate the following: fourteen less than three plus a number

Translate this problem as follows:

1. **Notice the turnaround word THAN; draw a box around the turnaround word.**

   fourteen less [than] three plus a number

   The two expressions to be turned around are "fourteen" and "three plus a number."

2. **Because the second expression includes the keyword PLUS, that expression must be enclosed in parentheses.**

   fourteen less [than] (three plus a number)

3. **Translate each expression and turn them around.**

   The expression translates to $(3 + x) - 14$.

Note: If the expression had included a comma between "three" and PLUS, the expression would have translated to $(3 - 14) + x$.

Chapter Checkout

Q&A

Translate the following into algebraic expressions.

1. Nineteen minus six increased by two
2. The sum of fourteen and a number, subtracted from five
3. The difference of five and a number, subtracted from four

4. The product of five and the sum of a number and five
5. Subtract three from the sum of a number and four
6. Add the difference between a number and four to twice the number
7. Divide eight by the sum of a number and four
8. The sum of one and the quotient of fifty-six and eight
9. The difference between the sum of a number and five and a number
10. Two times the quotient of a number and eight
11. Ten minus the sum of three and four
12. Three times the sum of a number and four
13. Nine added to a number minus five

Answers: 1. $19 - 6 + 2$ **2.** $5 - (14 + x)$ **3.** $4 - (5 - x)$ **4.** $5(x + 5)$ **5.** $(x + 4) - 3$ **6.** $(2x) + (x - 4)$ **7.** $8 \div (x + 4)$ **8.** $1 + (56 \div 8)$ **9.** $(x + 5) - x$ **10.** $2(x \div 8)$ **11.** $10 - (3 + 4)$ **12.** $3(x + 4)$ **13.** $(x - 5) + 9$

Chapter 3
SIMPLIFYING EXPRESSIONS

Chapter Check-In

- ❑ Distributive property of multiplication over addition
- ❑ Combining like terms
- ❑ Distributing a negative number

Chapters 1 and 2 explain how to translate expressions. Simplifying those expressions is the next step and is the focus of this chapter. An algebraic expression is simplified by using the **distributive property** and combining **like terms.**

Use Example 1 to test your translating skills. While reading the material, you may want to cover up the answer to the example to see whether you can arrive at the same answer on your own. This method lets you review what you have learned up to this point. If you need help determining which keywords are leading keywords, what the turnaround words are, and when to include parentheses, refer to Chapters 1 and 2, as well as the Pocket Guide at the front of this book.

Example 1: Translate the following: six added to four times the difference between a number and three

Read the problem carefully, looking for any keywords that indicate the operation, and then follow these steps:

1. **When you see a turnaround word (TO, FROM, or THAN), draw a box around the turnaround word to help you remember to change the order of the translation.**

   In this case, you have the turnaround word TO.

   six added [to] four times the difference
   between a number and three

2. **When you see two adjacent keywords, separate the two keywords with an open parenthesis and a closed parenthesis at the end of the expression.**

   Here, the keywords TIMES and DIFFERENCE BETWEEN are adjacent. Do not be thrown off when the word "the" or "a" appears between adjacent keywords; the keywords are still considered adjacent.

   six added [to] four times (the difference between a number and three)

3. **Look for a leading keyword; when you see one, circle the leading keyword and indicate the keyword to which it leads, underlining the expressions before and after the corresponding keyword.**

   Here, DIFFERENCE BETWEEN is the leading keyword, and you underline the expressions before and after the corresponding keyword AND. Use an arrow to indicate that the subtraction sign goes between "a number" and "three."

   six added [to] four times (the (difference) between a number and three)

   The expression translates to $4(x - 3) + 6$.

Distributive Property of Multiplication over Addition

When you see parentheses in an expression, use the distributive property to remove the parentheses before combining like terms, as shown in Example 2.

Example 2: Simplify the following expression: $5(2x + 7)$

The distributive property allows you to distribute the multiplication by five to each **term** in the parentheses. Drawing arrows to each term may help you remember to multiply every term.

5 times $(2x + 7)$

When a number is placed adjacent to a parentheses or a variable, multiplication is implied. In Example 2, multiplication is implied between the

5 and the expression in parentheses. (Multiplication is also implied between the 2 and the x, but that expression is simplified, so no further simplification needs to be done to it.) To simplify, multiply 5 times each term in the parentheses. The expression in simplified form is $10x + 35$.

Although this could also be written as $35 + 10x$, the standard in mathematics is that a term containing variables (such as x) comes first, while constant terms (terms without variables) come last.

Example 3: Simplify the following expression using the distributive property: $6(x + 5)$

Distribute the 6 over the terms in the parentheses; in other words, multiply each term in the parentheses by 6.

$$6 \text{ times } (x + 5)$$

After using the distributive property, the expression in simplified form is $6x + 30$.

Example 4: Use the distributive property to simplify the expression: $2(3x + y + 4)$

Distribute the 2 to all three terms in the parentheses.

$$2 \text{ times } (3x + y + 4)$$

The expression in simplified form is $6x + 2y + 8$

Combining Like Terms

Like terms are terms with the same variables raised to the same powers, as shown in Example 5.

Example 5: Simplify the following expression: $9x^2 + 3x^2 - 4y$

The first two terms are like terms because they both have a constant multiplied by the variable x raised to the second power. Like terms can be added or subtracted, like this: $12x^2 - 4y$

You have probably heard the expression "you can't add apples and oranges." Combining like terms can be thought of using that analogy. If you read Example 5 as "nine apples plus three apples minus four oranges," you can see that you can add apples to apples (like terms). Twelve apples minus four oranges is as simple as the phrase can get. You cannot subtract four from twelve and get eight because the units would not make sense.

Example 6: Simplify the expression translated in Example 1 of this chapter: $4(x - 3) + 6$

Follow these steps to solve Example 6:

1. **Distribute the 4 to each of the terms within the parentheses.**

   $$4 \text{ times } (x - 3) + 6$$

   After multiplying, the expression becomes $4x - 12 + 6$

2. **Look for like terms and combine them.**

   Are there any like terms? Yes: -12 and 6 are both constant terms and are, therefore, like terms. When you add -12 and 6, the result is -6.

   The simplified equation is $4x - 6$.

   Remember: If you need to review operations with signed numbers, read CliffsQuickReview *Basic Math and Pre-Algebra* by Jerry Bobrow (Wiley Publishing).

Example 7 uses many of the concepts covered to this point.

Example 7: Translate and simplify the following: twice the sum of five and a number is added to twelve

Carefully read the expression and notice:

The two adjacent key words: TWICE the SUM OF

The separator: IS

The leading keyword: the SUM OF ______ AND ______

The turnaround word: TO

Here's how to proceed:

1. **Place opening and closing parentheses, as determined by the adjacent keywords, TWICE and SUM OF, and the separator, IS.**

   twice (the sum of five and a number) is added to twelve

2. **Within the parentheses is the leading keyword SUM OF. Underline the two expressions to be added.**

   twice (the sum of five and a number) is added to twelve

3. **The leading keyword SUM OF indicates a plus sign where the corresponding keyword AND is in the expression. Draw an arrow to indicate the placement of the plus sign.**

   twice (the sum of five and
   a number) is added to twelve

4. **The turnaround word TO indicates the twelve will be at the beginning of the expression instead of the end. Place a box around TO.**

   twice (the sum of five and
   a number) is added to twelve

5. **The expression translates to 12 + 2(5 + x).**

   The expression is now translated, but it needs to be simplified.

6. **When you see parentheses in an expression, use the distributive property to remove the parentheses before combining like terms.**

   Here, multiply 2 times each term in the parentheses.

   $12 + 2 \text{ times } (5 + x)$

   This gives you $12 + 10 + 2x$.

   Note: If you try to combine 12 and 2 before distributing, you arrive at an incorrect answer. Always use the distributive property to get rid of the parentheses *before* combining like terms.

7. **Combine like terms, and you get 22 + 2x.**

8. **Put the equation in the standard form (variables first), and your simplified equation is 2x + 22.**

Distributing a Negative Number

Pay special attention when distributing a negative number.

Example 8: Simplify the following: $-2(x + 3) - 6$

You can simplify the expression as follows:

1. **Multiply both terms in the parentheses by –2.**

$$-2 \text{ times } (x + 3) - 6$$

This gives you $-2x - 6 - 6$

Remember: When you multiply a negative number by a positive number, the answer is always negative. When both numbers have the same sign (whether positive or negative), the answer is always positive.

2. **Now combine like terms: –2x – 12.**

   This is the simplified expression.

While a problem such as Example 8 does not seem difficult, if the order is changed slightly, the distribution is not as obvious, as shown in Example 9.

Example 9: Simplify the following: $-7 - 2(x + 3)$

A common error is to distribute a +2. A –2 must be distributed throughout the parentheses because the entire expression is subtracted from –7. If you distribute a +2, the expression will be incorrect because you will add 6 instead of subtracting 6.

Remember that subtracting 2 yields the same result as adding –2. When subtracting signed numbers, you may want to change the subtraction sign to an addition sign and change the sign of the second number, like this: $-7 + -2(x + 3)$.

You then simplify the equation as follows:

1. **Distribute –2 to each term in the parentheses, like this:**

$$-7 + -2 \text{ times } (x + 3)$$

2. **Simplify the equation to –7 + –2x + –6.**
3. **Using commutative property of addition, place like terms adjacent to each other: –2x + –7 + –6.**
4. **Combine like terms to yield the simplified answer: –2x – 13.**

Sometimes, you are given a negative sign, but no number has to be distributed over the parentheses. In this case, you can use the **identity property of multiplication** to place a 1 in front of the parentheses so that you can distribute a –1. Remember that any number multiplied by 1 has the same value.

Example 10: Simplify the following expression: $-17 - (x - 12)$

In this expression, $x - 12$ is being subtracted from -17. Follow these steps:

1. **Use the identity property of multiplication to multiply the expression in parentheses by 1, like this: $-17 - 1(x - 12)$.**
2. **Change subtracting 1 to adding a -1: $-17 + -1(x - 12)$.**
3. **Distribute the negative one: $-17 + -1x + 12$.**
4. **Using commutative property of addition, place like terms adjacent to each other: $-1x + -17 + 12$.**
5. **Combine like terms to yield the expression in simplified form: $-1x - 5$.**
6. **Mathematics standards dictate that you do not have to include coefficients of 1, so you may simplify this one more time to $-x - 5$.**

Chapter Checkout

Q&A

Simplify the following expressions by using the distributive property and combining like terms.

1. $3x + 2(x + 2)$
2. $x + 5(x + 4)$
3. $4x + 3(2x + 6)$
4. $10 + 2(x - 3)$
5. $3(x - 3) + 6$
6. $2(x + 5) - 3x$
7. $-3(x + 4) + 5$
8. $-2(x - 4) + 5$
9. $3x - 2(x + 6)$
10. $4x - 2(x - 2)$
11. $-3(x - 5) + 6$
12. $x - 2(x + 4)$
13. $4 - (x - 5)$
14. $3 - (x + 4)$
15. $9 - (x + y - 2)$

Answers: **1.** $5x + 4$ **2.** $6x + 20$ **3.** $10x + 18$ **4.** $2x + 4$ **5.** $3x - 3$ **6.** $-x + 10$ **7.** $-3x - 7$ **8.** $-2x + 13$ **9.** $x - 12$ **10.** $2x + 4$ **11.** $-3x + 21$ **12.** $-x - 8$ **13.** $-x + 9$ **14.** $-x - 1$ **15.** $-x - y + 11$

Chapter 4
EQUATIONS

Chapter Check-In

- ❑ Keywords indicating equality
- ❑ Solving simple linear equations
- ❑ Checking solutions
- ❑ Checking translations

Translating English statements into equations is the next step in the successful solving of word problems. An **equation** is made up of two expressions that are set equal to each other. The easiest way to differentiate between an expression and an equation is that an equation has an equal sign.

In this chapter, you solve equations, check your solutions and your translations, and continue translating and simplifying expressions (see Chapters 1 through 3).

Keywords Indicating Equality

One of the hardest parts of solving word problems is creating the correct equation after reading the problem. Many people agree that, in comparison to setting up the correct equation, solving the equation is easy.

The following keywords indicate equality and, when translated, become the equal symbol, =.

IS
IS EQUAL TO
EQUALS
YIELDS

RESULTS IN
THE RESULT IS
IS THE SAME AS

The **direct translation strategy** works for all the equality keywords, which means you do not need to change the order. The equal sign is placed exactly where the keyword is located.

Example 1: Translate the following sentence into an algebraic equation: Twice the difference between a number and five is equal to negative fourteen

Start by making the appropriate markings on the English sentence as you read it. Did you notice the adjacent keywords, the leading keywords, and the IS? Here's how you make the appropriate markings:

1. **Put an open parenthesis between the adjacent keywords and close the parentheses before IS.**

   Twice (the difference between a number and five) is equal to negative fourteen.

2. **Underline the two expressions to be subtracted as indicated by the leading keyword, DIFFERENCE BETWEEN.**

   Twice (the difference between a number and five) is equal to negative fourteen.

3. **Use an arrow to remind you that AND is replaced by a minus sign.**

   Twice (the difference between a number and

   five) is equal to negative fourteen.

   Note: The TO that is part of IS EQUAL TO is not a turnaround word. Remember that the direct translation strategy works for all the equality keywords. The expression on the left side of the equation does not have to be turned around with the expression on the right side.

4. **Translate the equation.**

The equation translates to $2(x - 5) = -14$.

In the following section, you not only translate but also solve. For now, though, just practice translating.

One of the most confusing keywords indicating equality is IS. In a translation problem, you may see more than one IS, as Example 2 demonstrates.

Example 2: Translate the following sentences into an algebraic equation: When a number is subtracted from twelve, the result is five. What is the number?

The second IS is the only one translated to the = symbol. Any IS placed directly in front of a keyword for an operation does not indicate equality. When a number directly follows IS, however, IS does indicate equality.

Remember: Mark the turnaround word to remind you to translate in the correct order.

When a number is subtracted from
twelve, the result is five. What is the number?

The equation translates to $12 - x = 5$.

Check more of your translation abilities with Example 3.

Example 3: Translate the following statement into an equation: The product of eight and a number yields twenty-four.

What operation should replace AND in this equation? The leading keyword, PRODUCT OF, indicates the eight is multiplied by a number.

The product of eight and a number

yields twenty-four

Note: The first expression is translated, then the equal sign is translated, and then the expression on the right of the equal sign. Often, in mathematics, breaking the problem up into smaller pieces can increase your success. Instead of translating the whole equation at one time, translate the first expression, the equal sign, and then the second expression.

The equation translates to $8x = 24$.

Solving Simple Linear Equations

Algebraic equations are translated from complete English sentences. In Chapter 1, you translate English phrases to algebraic expressions, which cannot be solved; they can only be **simplified.** Equations, however, can be solved. In fact, in order to successfully solve a word problem, an equation must be written and solved.

Look at these two definitions in the following sections and compare the examples to ensure you know the distinction between an expression and an equation.

Defining an algebraic expression

An **algebraic expression** is a collection of constants, variables, symbols of operations, and grouping symbols, as shown in Example 4.

Example 4: $4(x - 3) + 6$

Defining an algebraic equation

An algebraic equation is a statement that two algebraic expressions are equal, as shown in Example 5.

Example 5: $4(x - 3) + 6 = 14 + 2x$

The easiest way to distinguish a math problem as an equation is to notice an equals sign.

In Example 6, you take the algebraic expression given in Example 4 and simplify it to review the process of simplification. An algebraic expression is simplified by using the **distributive property** and combining **like terms.**

Example 6: Simplify the following expression: $4(x - 3) + 6$

Here is how you simplify this expression:

1. **Remove the parentheses using the distributive property.**

   $4x + -12 + 6$

2. **Combine like terms.**

   The simplified expression is $4x + -6$.

Note: This problem does not solve for *x*. This is because the original problem is an expression, not an equation, and, therefore, cannot be solved.

Four steps for solving simple linear equations

In order to solve an equation, follow these steps:

1. **Simplify both sides of the equation by using the distributive property and combining like terms, if possible.**
2. **Move all terms with variables to one side of the equation using the addition property of equations, and then simplify.**

3. **Move the constants to the other side of the equation using the addition property of equations and simplify.**
4. **Divide by the coefficient using the multiplication property of equations.**

In Example 7, you solve the equation given in Example 5, using the four preceding steps to find the solution to the equation.

Example 7: Solve the following equation: $4(x - 3) + 6 = 14 + 2x$

Use the four steps to solving a linear equation, as follows:

1. **Distribute and combine like terms.**

$$4x - 12 + 6 = 14 + 2x$$
$$4x - 6 = 14 + 2x$$

2a. Move all terms with variables to the left side of the equation.

In this example, add a **–2x** to each side of the equation.

$$4x - 6 + \mathbf{-2x} = 14 + 2x + \mathbf{-2x}$$

The addition property of equations states that if the same term is added to both sides of the equation, the equation remains a true statement. The addition property of equations also holds true for subtracting the same term from both sides of the equation.

2b. Place like terms adjacent to each other and simplify.

Note: Subtracting 6 is changed to adding –6 because the commutative property of addition works only if all operations are addition.

$$4x + -2x + -6 = 14$$
$$2x - 6 = 14$$

3. **Move the constants to the right side of the equation and simplify.**

$$2x - 6 + \mathbf{6} = 14 + \mathbf{6}$$

Note: The opposite operation was used to move the constant.

$$2x = 20$$

4. **Divide by the coefficient and simplify.**

$$\frac{2x}{2} = \frac{20}{2}$$

The solution is $x = 10$.

Example 8: Solve the following equation: $12 + 2(3x - 7) = 5x - 4$

Use the four steps to solving a linear equation, as follows:

1a. Distribute and combine like terms.

$$12 + 6x - 14 = 5x - 4$$

1b. Place like terms adjacent to each other and simplify.

$$6x + 12 + -14 = 5x - 4$$
$$6x - 2 = 5x - 4$$

2a. Move variables to the left side of the equation.

In this example, add $-5x$ to each side of the equation.

$$6x - 2 + \mathbf{-5x} = 5x - 4 + \mathbf{-5x}$$

2b. Place like terms adjacent to each other and simplify.

Note: All subtractions are changed to addition of a negative number.

$$6x + -5x + -2 = 5x + -5x + -4$$
$$1x - 2 = -4$$

3. Move the constants to the right side of the equation and simplify.

$$x - 2 + \mathbf{2} = -4 + \mathbf{2}$$

Note: The opposite operation was used to move the constant.

4. Because the coefficient is 1, Step 4 is not necessary.

The solution is $x = -2$.

Example 9: Solve the following equation: $6 - 3(2 - x) = -5x + 40$

Use the four steps to solving a linear equation, as follows:

1. Distribute and combine like terms.

$$6 - 6 + 3x = -5x + 40$$

Did you remember to distribute the negative three?

$$3x = -5x + 40$$

2a. Move variables to the left side of the equation.

In this example, add $5x$ to each side of the equation.

$$3x + \mathbf{5x} = -5x + 40 + \mathbf{5x}$$

2b. Place like terms adjacent to each other.

$$3x + 5x = -5x + 5x + 40$$

2c. Simplify by combining like terms.

$$8x = 40$$

3. This step is not necessary in this example because all of the constants are on the right side of the equation.

4. Divide by the coefficient and simplify.

$$\frac{8x}{8} = \frac{40}{8}$$

The solution is $x = 5$.

Remember: The four steps for solving equations must be done in order, but not all steps are necessary in every problem.

Checking Solutions and Translations

Confidence in your mathematical skills increases when you check your solution. Your solution can be checked two ways: in the equation and in the original word problem.

First, solve an equation in Example 10 and check the solution in Example 11.

Example 10: Solve the following equation: $3(2x - 1) = 4x + 1$

1. Distribute.

$$6x - 3 = 4x + 1$$

There was no need to combine like terms because there are no like terms in each expression.

2a. Move variables to the left side of the equation.

In this example, add $-4x$ to each side of the equation.

$$6x - 3 + \mathbf{-4x} = 4x + 1 + \mathbf{-4x}$$

2b. Place like terms adjacent to each other and simplify.

Note: Subtracting 3 is changed to adding –3. The commutative property of addition works only if all operations are addition.

$$6x + -4x + -3 = 4x + -4x + 1$$
$$2x - 3 = 1$$

3. Move the constants to the right side of the equation and simplify.

$$2x - 3 + \mathbf{3} = 1 + \mathbf{3}$$

Note: The opposite operation was used to move the constant.

$$2x = 4$$

4. **Divide by the coefficient and simplify.**

$$\frac{2x}{2} = \frac{4}{2}$$

The solution is $x = 2$.

Checking solutions

To check the solution to an equation, replace all of the variables in the original problem with your solution. Evaluate both sides of the equation as if they were **order of operations** problems. For a review of order of operations see *CliffsQuickReview Basic Math and Pre-Algebra* by Jerry Bobrow (Wiley Publishing).

Example 11: Check the solution to Example 10.

1. **Replace the variables in the original equation and place parentheses around your solution to prevent sign errors.**

$$3[2(2) - 1] = 4(2) + 1$$

2. **Evaluate the expression in the parentheses first, performing one operation at a time.**

   Remember: The brackets [ ] have the same meaning as parentheses ( ). Even though this example has multiple parentheses, only the set of brackets contains operations. Those operations must be performed first.

   Inside the brackets are two operations, implied multiplication and subtraction. When following the order of operations, multiplication is performed before subtraction.

$$3(4 - 1) = 4(2) + 1$$
$$3(3) = 4(2) + 1$$

3. **Evaluate any exponents.**

   Because there are no exponents, proceed to the following step.

4. **Perform all multiplication and division, working from left to right.**

$$9 = 8 + 1$$

Note only one operation was done in each of the two expressions. To be successful with the order of operations within each expression, each operation must be performed one at a time.

5. **Perform any addition and subtraction, one operation at a time.**

$$9 = 9 \checkmark$$

The solution is correct if both numbers are exactly the same.

In Example 12, you check the solution of the equation solved in Example 7.

Example 12: Check the solution $x = 10$ of the equation $4(x - 3) + 6 = 14 + 2x$

1. **Replace the variables in the original equation and place parentheses around your solution.**

$$4[(10) - 3] + 6 = 14 + 2(10)$$

2. **Evaluate the expression in the parentheses first, performing one operation at a time.**

Even though there are multiple parentheses in this example, subtraction is the only operation in parentheses. Subtraction must be performed first.

$$4(7) + 6 = 14 + 2(10)$$

No **evaluation** was done on the right side of the equation because there were no operations inside of the parentheses.

3. **Evaluate any exponents.**

Because there are no exponents, move to the next step.

4. **Perform all multiplication and division, working from left to right.**

$$28 + 6 = 14 + 20$$

Note only one operation was done on each side of the equation. To be successful with the order of operations, perform one operation at a time.

5. **Perform any addition and subtraction, one operation at a time.**

$$34 = 34 \checkmark$$

Because both expressions evaluated to be 34, the solution checks.

Example 13 translates a word problem from start to finish, from translating to checking.

Example 13: Translate, solve, and check the following word problem: Twice the difference between three and a number is equal to four times the number. What is the number?

Here is how you solve this problem:

1. **Make the markings that are helpful when you have adjacent keywords, a separator, a leading keyword, and/or a turnaround word.**

   Twice (the difference between three and a number) is equal to four times the number. What is the number?

   Twice (the difference between

   three and a number) is equal to four

   times the number. What is the number?

2. **Translate the words into the correct symbols.**

   $$2(3 - x) = 4x$$

3. **Distribute and combine like terms.**

   $$6 - 2x = 4x$$

4. **Move the variables to the left side of the equation.**

   $$6 - 2x - \mathbf{4x} = 4x - \mathbf{4x}$$

5. **Simplify each side of the equation.**

   $$6 - 6x = 0$$

   Note: The right side of the equation is now 0. The equation must have expressions on either side of the equality sign to be solvable. You cannot just leave the right side of the equation blank. 0 is the expression on the right side of the equation.

6. **Move the constant to the right side of the equation.**

   Remember: Be careful with negative signs.

   $$6 - 6x - \mathbf{6} = 0 - \mathbf{6}$$

7. **Place like terms adjacent to each other.**

   $$-6x + 6 + -6 = 0 - 6$$

8. **Simplify the expressions on each side of the equation.**

$$-6x = -6$$

9. **Divide by the coefficient.**

$$\frac{-6x}{-6} = \frac{-6}{-6}$$

The solution is $x = 1$

10. **To check the solution, replace the variables in the original equation and place parentheses around your solution.**

$$2(3 - 1) = 4(1)$$

11. **Follow the order of operations to simplify the expressions on each side of the equation.**

$$2(2) = 4(1)$$
$$4 = 4 \checkmark$$

Checking your translation

Not only can you check your solving abilities, you can also check your translating abilities. Check your translation by rereading the original translation word problem and replacing the word "number" with 1, which is the solution.

The original problem reads: Twice the difference between three and a number is equal to four times the number. Every place you see the word "number" replace it with 1.

> Twice the difference between three and 1 (pause here because the IS acts as a separator)

What is the difference between three and 1? The difference between three and 1 is 2 and twice 2 is 4. So, this expression evaluates as 4.

Evaluate the rest of the statement, "is equal to four times the number." Replace the number with 1. Four times 1 is also 4. Both your solution and your translation check using mental math.

Example 14 is a good one to use to test yourself. (**Remember:** Successful students test their own skills by trying to work a problem on their own before looking in the book at a solution.)

Example 14: Translate, solve, and check the following word problem: Five plus two times a number equals seventeen. Find the number.

1. **Make any markings that may be helpful for adjacent keywords, a separator, a leading keyword, and/or a turnaround word.**

   These markings are not necessary in this problem.

2. **Translate the words into the correct symbols.**

   $$5 + 2x = 17$$

3. **Distribute and combine like terms.**

   In this example, there are no parentheses and no like terms on each side of the equation.

4. **Move the variables to the left side of the equation.**

   All of the variables are already on the left side of the equation.

5. **Move the constant to the right side of the equation using the addition property of equations.**

   $$5 + 2x + -5 = 17 + -5$$

6. **Place like terms adjacent to each other.**

   $$2x + 5 + -5 = 17 -5$$

7. **Combine like terms.**

   $$2x = 12$$

8. **Divide by the coefficient.**

   $$\frac{2x}{2} = \frac{12}{2}$$

   The solution is $x = 6$

9. **Check the solution by replacing all of the variables in the original problem with your solution.**

   $$5 + 2(6) = 17$$

10. **Use the order of operations to evaluate each expression on either side of the equation.**

    Multiplication is the first operation to perform.

    $$5 + 12 = 17$$

    Addition is the second operation to perform.

    $$17 = 17 \checkmark$$

11. **Check your translation by rereading the original translation word problem and replacing the word "number" with 6, which is the solution.**

five plus two times six (pause here because EQUALS separates the two expressions)

Think "two times six" is twelve and "five plus twelve" is seventeen, which is equal to the expression on the right side of the equation.

Both your solution and your translation check using mental math.

Chapter Checkout

Q&A

Translate and solve the following equations. Be sure to check your solution and translation.

1. Twice the difference between four and a number is sixteen. Find the number.
2. The sum of five and a number is three. Find the number.
3. The difference between two times a number and four is eight. Find the number.
4. Three times a number added to eight is fourteen. Find the number.
5. A number plus the quotient of fifty-six and seven is eleven. Find the number.
6. Three times the difference between three and a number is eighteen. Find the number.
7. Two times a number is the same as the difference between the number and four. What is the number?
8. Four times a number plus three times the sum of a number and four is equal to negative nine. Find the number.
9. Three times the total of a number and six is thirty. Find the number.
10. The product of five and a number is twenty-five. Find the number.
11. Five less than three times a number is sixteen. Find the number.
12. Three more than twice a number is thirteen. Find the number.
13. The difference between a number and two times the number is nine. Find the number.
14. The sum of a number and three times the number is fourteen. Find the number.
15. Three more than five times a number is twenty-three. Find the number.

Answers: 1. $2(4 - x) = 16$; $x = -4$ **2.** $5 + x = 3$; $x = -2$ **3.** $(2x) - 4 = 8$; $x = 6$ **4.** $8 + 3x = 14$; $x = 2$ **5.** $x + 56 \div 7 = 11$; $x = 3$ **6.** $3(3 - x) = 18$; $x = -3$ **7.** $2x = x - 4$; $x = -4$ **8.** $4x + 3(x + 4) = -9$; $x = -3$ **9.** $3(x + 6) = 30$; $x = 4$ **10.** $5x = 25$; $x = 5$ **11.** $3x - 5 = 16$; $x = 7$ **12.** $2x + 3 = 13$; $x = 5$ **13.** $x - 2x = 9$; $x = -9$ **14.** $x + 3x = 14$; $x = \frac{7}{2}$ **15.** $5x + 3 = 23$; $x = 4$

Mixed Review

Translate the following. Then, identify the problem as an expression or an equation: If it is an expression, simplify the expression; and if it is an equation, translate, solve, and check the equation.

1. The difference between a number and five equals twenty-four. Find the number.
2. Five times the sum of a number and two.
3. The sum of a number and three, subtracted from the difference between the number and two.
4. Seven multiplied by the difference between a number and two.
5. The difference between a number and fifteen is six. Find the number.
6. Twice the sum of five and a number is added to ten.
7. Two times the sum of a number and four is subtracted from three.
8. Three times the sum of a number and four is equal to four times the difference between two times the number and five. What is the number?
9. The total of twelve and a number is equal to negative twelve. Find the number.
10. One third of seventy-five subtracted from fifty.
11. Three more than twice a number, added to ten.
12. Four times the sum of a number and three is added to the difference between a number and two.
13. Three times a number increased by eight.
14. Five times a number yields twenty-five. What is the number?
15. Add two times the sum of a number and four to six.
16. The product of four and a number plus one is eight. Find the number.
17. A number subtracted from the sum of a number and five.

18. The sum of a number and three is subtracted from the difference between a number and two.
19. The sum of six and a number is subtracted from five.
20. Four times the sum of a number and three is added to five.
21. The product of six and seven is subtracted from forty-nine.
22. The difference between a number and two is added to three times the difference between four and the number.
23. Add a number to seven.
24. The sum of three and a number is two. Find the number.
25. The product of five and a number is negative forty-five. Find the number.
26. A number subtracted from the product of six and a number.
27. Three less than five times a number is seven. Find the number.
28. Three times the sum of twice a number and one is the same as negative nine. Find the number.
29. A number is two times the difference between eighteen and the number. Find the number.
30. Find a number such that two times the number is equal to twenty-eight.
31. Ten more than a number is the same as twice the number. Find the number.
32. The difference between four times a number and twice the number is twelve. Find the number.

Answers: 1. equation; $x - 5 = 24$; $x = 29$ **2.** expression; $5(x + 2)$; $5x + 10$ **3.** expression; $(x - 2) - (x + 3)$; -5 **4.** expression; $7(x - 2)$; $7x - 14$ **5.** equation; $x - 15 = 6$; $x = 21$ **6.** expression; $10 + 2(5 + x)$; $2x + 20$ **7.** expression; $3 - 2(x - 4)$; $-2x - 5$ **8.** equation; $3(x+4) = 4(2x-5)$; $x = \frac{32}{5}$ **9.** equation; $12 + x = -12$; $x = 24$ **10.** expression; $50 - \frac{1}{3}(75)$; 25- **11.** expression; $10 + (2x + 3)$; $2x + 13$ **12.** expression; $(x - 2) + 4(x + 3)$; $5x + 10$ **13.** expression; $3x + 8$; $3x + 8$ **14.** equation; $5x = 25$; $x = 5$ **15.** expression; $6 + 2(x + 4)$; $2x + 14$ **16.** equation; $4(x + 1) = 8$; $x = 1$ **17.** expression; $(x + 5) - x$; 5 **18.** expression; $(x - 2) - (x + 3)$; -5 **19.** expression; $5 - (6 + x)$; $-x - 1$ **20.** expression;

$5 + 4(x + 3)$; $4x + 17$ **21.** expression; $49 - (6 \times 7)$; 7 **22.** expression; $3(4 - x) + (x - 2)$; $-2x + 10$ **23.** expression; $7 + x$; $7 + x$ **24.** equation; $3 + x = 2$; $x = -1$ **25.** equation; $5x = -45$; $x = -9$ **26.** expression; $6x - x$; $5x$ **27.** equation; $5x - 3 = 7$; $x = 2$ **28.** equation; $3(2x + 1) = -9$; $x = -2$ **29.** equation; $x = 2(18 - x)$; $x = 12$ **30.** equation; $2x = 28$; $x = 14$ **31.** equation; $x + 10 = 2x$; $x = 10$ **32.** equation; $4x - 2x = 12$; $x = 6$

Chapter 5
GEOMETRY PROBLEMS

Chapter Check-In

- ❑ Triangles
- ❑ Rectangles and squares
- ❑ Multiple shapes
- ❑ Other easily visualized word problems

Geometry word problems are easy to visualize and, consequently, easy to understand.

Triangles

A **triangle** is a three-sided closed figure. The **vertices** are used to identify the triangle and also identify the three angles in the triangle.

Some triangle word problems include terminology and symbols describing the relationships of the measures of angles and lengths of sides. A brief review of symbols is given in this section. Definitions of **acute, obtuse, equilateral,** and **scalene triangles** are in the glossary.

A **right triangle** is a triangle with one **right angle,** an angle that measures 90°, as shown in Figure 5-1.

Figure 5-1 The right triangle ΔABC.

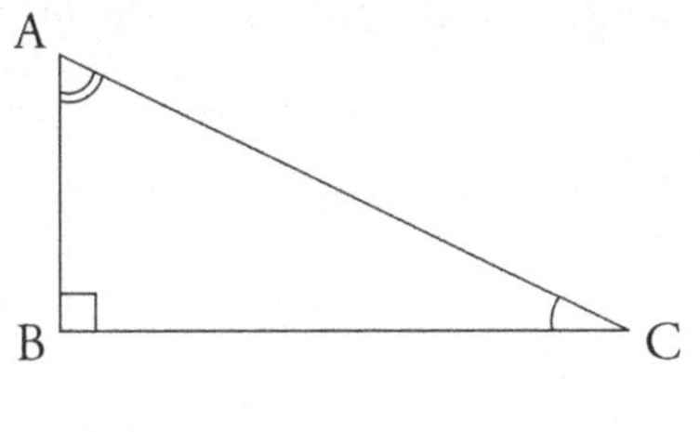

The angle that measures 90° can be shown with symbols. One symbol used to show the measure of an angle is $m\angle ABC = 90°$. The square on the angle ABC is another symbol indicating the angle measures 90°.

Note: A standard way to reference an angle is with the three letters on the vertices. For example, the right angle is named $\angle ABC$ with the vertex B as the middle letter.

The right triangle in Figure 5-1 has three angles with different measures as indicated by the arc symbols in each angle. If the same arc is in two angles, those angles have the same measure.

In order to work word problems involving triangles, you should know some basic facts about triangles, as discussed in the following sections.

Perimeter of a triangle

The **perimeter** of a closed figure is the sum of the lengths of its sides, as illustrated in Figure 5-2. The hash marks on each side indicate that all three sides are of different length in this particular triangle. The formula for perimeter of a triangle is $P = a + b + c$.

Figure 5-2 Triangle showing the lengths of the sides.

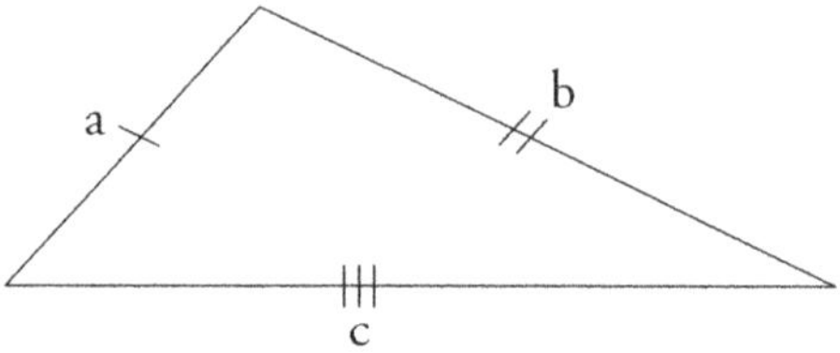

Example 1: Find the perimeter of a scalene triangle with sides of 2 m, 5 m, and 6 m.

To solve this problem, follow these steps:

1. **Choose the correct formula.**

$$P = a + b + c$$

2. **Substitute the lengths of the sides into the formula.**

$$P = 2 + 5 + 6$$

3. Solve for the perimeter one step at a time.

$$P = 7 + 6$$
$$P = 13$$
$$P = 13 \text{ m}$$

The perimeter of the triangle is thirteen meters.

Note: You must put units on your answer to geometric word problems. There is a large difference, for example, between 13 meters and 13 inches.

Remember: You can use two different ways to correctly solve Example 1. You can use the formula for the perimeter of a triangle, $P = a + b + c$, substitute the given values for the variables, and then solve the equation for the one unknown variable. You also could use the broad definition of perimeter and simply add the lengths of the sides. Using the formula (an equation) is the preferred method. An equation is necessary to solve more complicated word problems.

Area of a triangle

The **area** of a figure is the amount of surface enclosed by a closed figure. One way to recognize an area word problem is to visualize an area rug covering the shape.

To find the area of a triangle use the formula $A = \frac{1}{2}bh$, where b represents the **base** of the triangle and h represents the **height.** The base and the height must be **perpendicular** to each other. This is represented in Figure 5-3 with a square on the angle formed by the intersection of the base and the height.

Figure 5-3 Triangle showing the base and the height.

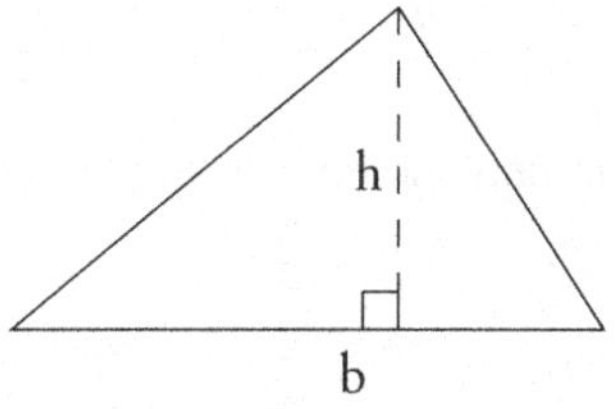

Example 2: Find the length of the base of a triangle with an area of 24 in^2 and a height of 6 inches.

Solve Example 2 in the following way:

1. **Choose the correct formula.**

$$A = \frac{1}{2}bh$$

2. **Substitute the given values for the variables into the formula.**

$$24 = \frac{1}{2}b(6)$$

3. **Using the commutative property of multiplication, place the $\frac{1}{2}$ adjacent to the 6 and multiply.**

$$24 = \frac{1}{2}(6)b$$

$$24 = 3b$$

4. **To solve for b, the base, divide both sides of the equation by the coefficient 3 and simplify.**

$$\frac{24}{3} = \frac{3b}{3}$$

$$8 = b$$

$$b = 8 \text{ in}$$

The base of the triangle is 8 inches.

Note: Area is measured in square units, in^2 (inches squared), and base and height are measured in linear units, in (inches).

Sum of the measures of the angles

The sum of the measures of the angles in any triangle is always 180°, as shown by the following formula and in Figure 5-4:

$$m\angle 1 + m\angle 2 + m\angle 3 = 180°$$

Figure 5-4 Triangle showing another way to reference angles.

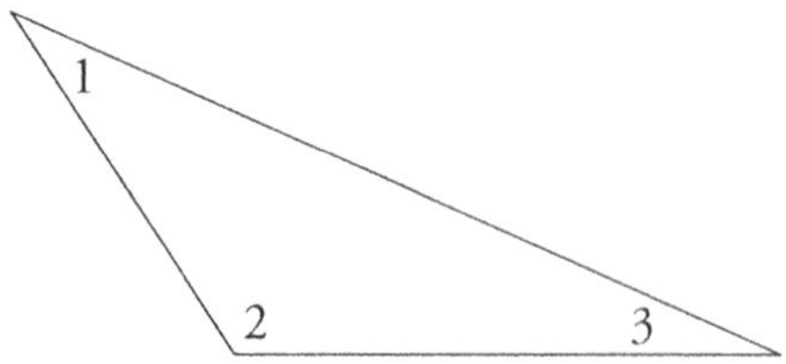

Example 3: Find the measure of ∠DEF, given the following diagram:

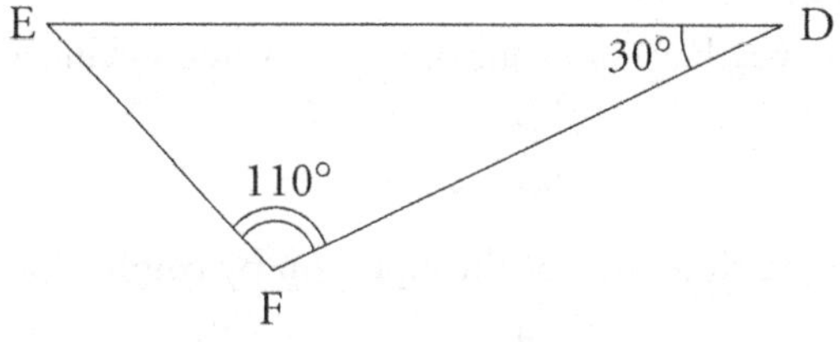

1. **From the figure, you can tell that m∠EFD = 110°; m∠EDF = 30°.**
2. **Using the formula for the sum of the angles of a triangle, substitute the given values.**

   You need to use only one unknown when solving an equation. Copy the formula.

   $$m\angle 1 + m\angle 2 + m\angle 3 = 180°$$

3. **Substitute the appropriate values.**

   $$m\angle \text{DEF} + 110 + 30 = 180$$

4. **Simplify the left side of the equation.**

   $$m\angle \text{DEF} + 140 = 180$$

5. **Isolate the unknown variable by subtracting 140 from both sides of the equation using the addition property of equations.**

   $$m\angle \text{DEF} + 140 - 140 = 180 - 140$$
   $$m\angle \text{DEF} = 40$$
   $$m\angle \text{DEF} = 40°$$

   The measure of the unknown angle is 40 degrees.

Note: The degree symbol (°) is the unit of measure.

Example 4: Find the length of the missing side of a triangle with a longest side that measures 20 cm and a shortest side that measures 9 cm. The perimeter of the triangle is 46 cm.

In this word problem, a diagram is not provided. To visualize the problem, draw your own diagram. You do not have to draw the diagram exactly to scale.

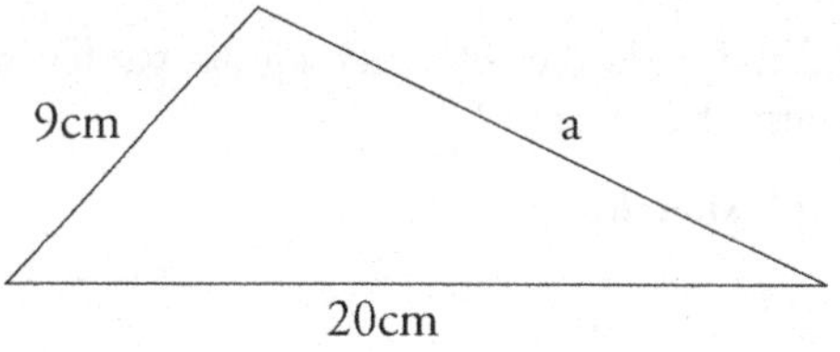

1. **Using the formula for perimeter of a triangle, substitute the given values.**

   Make sure you have only one unknown when solving an equation.

$$P = a + b + c$$
$$46 = a + 20 + 9$$

2. **Simplify the right side of the equation by combining like terms.**

$$46 = a + 29$$

3. **Isolate the unknown variable by subtracting 29 from both sides of the equation.**

$$46 - 29 = a + 29 - 29$$
$$17 = a$$
$$a = 17 \text{ cm}$$

   The measure of the unknown side is 17 centimeters.

Remember: As you complete a word problem, ask yourself two questions:

- Did I answer the question?
- Did I put units on my answer?

Rectangles and Squares

In order to work word problems involving rectangles and squares, you need to know their definitions and some basic facts.

- A **rectangle** is a four-sided closed figure in which the opposite sides are parallel and of equal length. All pairs of adjacent sides of a rectangle meet at right angles.
- A **square** is a special type of rectangle in which all sides are equal in length.

Perimeter of a rectangle

For a rectangle, the perimeter formula is $P = 2l + 2w$.

Example 5: Find the perimeter of a rectangular garden with a width of 15 feet and a length of 20 feet.

1. **Draw a rough sketch.**

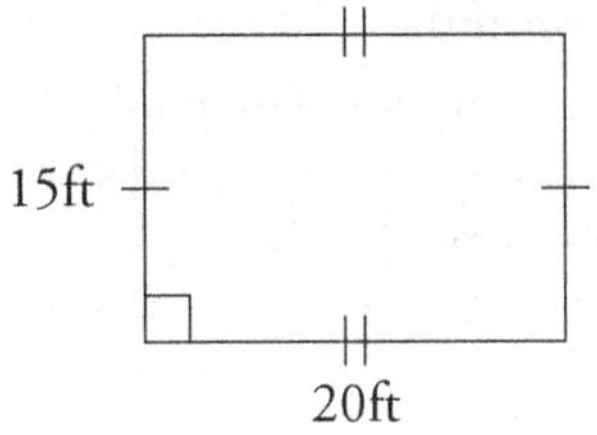

2. **Copy the correct formula.**

$$P = 2l + 2w$$

3. **Replace two of the variables with numbers from the problem.** This leaves only one unknown, the perimeter.

$$P = 2(20) + 2(15)$$

4. **Simplify using order of operations one step at a time.**

$$P = 40 + 2(15)$$

$$P = 40 + 30$$

$$P = 70$$

The perimeter of the rectangular garden is 70 feet.

Area of a rectangle

The formula for area of a rectangle is $A = l \times w$.

When measuring the area of any figure, the units for the area will be square units. In the formula for the area of a rectangle, length and width are multiplied. If length and width were both measured in yards, the product would be measured in yards squared.

Example 6: Sharon works at a carpet warehouse. When a customer wants to carpet his or her rectangular living room, Sharon measures the length and width of the living room. If the room measures 15 feet by 12 feet, what is the number of square feet of carpet to be purchased?

1. **After reading the problem carefully, draw a rough sketch of the living room.**

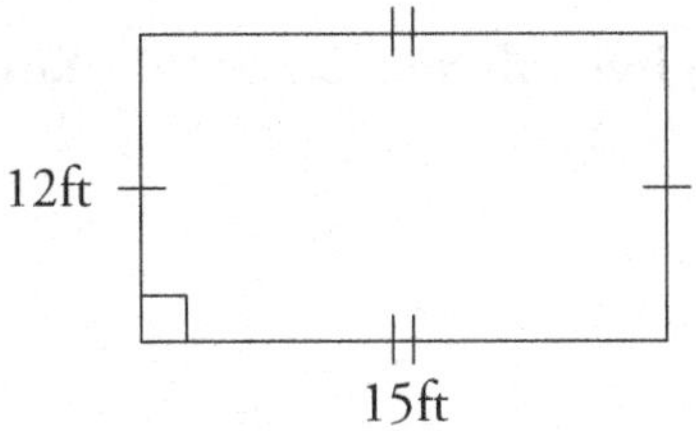

2. **Choose the right formula.**

   How can you tell that this problem is asking for area? The question in the word problem always identifies the unknown, and in this problem, the question asks for the number of square feet. The area formula is the only formula that uses square feet. The room is described as "rectangular"; therefore, use the formula for area of a rectangle.

$$A = l \times w$$

3. **Substitute the given information into the formula.**

$$A = 15 \times 12$$

4. **Multiply and solve the equation.**

$$A = 180$$

$$A = 180 \text{ ft}^2$$

   The number of square feet of carpeting needed for the living room is 180 square feet.

Perimeter of a square

The formula for perimeter of a square is $P = 4s$.

Example 7: If the perimeter of a square is 24 in, how long is each side?

1. **Draw a rough sketch.**

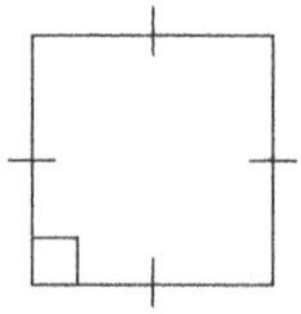

   Note: All sides are of equal length, as indicated by the single hash mark on each side. The square in the lower-left corner shows that all adjacent sides meet at right angles.

2. **Choose the applicable formula.**

$$P = 4s$$

3. **Substitute the given information into the formula.**

$$24 = 4s$$

4. Solve the equation by dividing both sides of the equation by four.

$$\frac{24}{4} = \frac{4s}{4}$$

$$6 = s$$

$$s = 6 \text{ in}$$

The length of each side of the square is 6 inches.

Area of a square

The formula for area of a square is $A = s^2$.

Example 8: Find the area of a square with sides of 6 ft.

1. Draw a rough sketch.

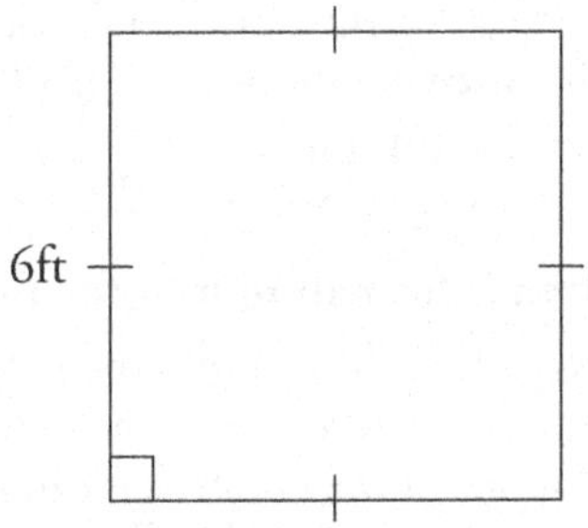

2. Choose the correct formula and plug in the correct numbers.

$$A = s^2$$

$$A = 6^2$$

$$A = 36$$

$$A = 36 \text{ ft}^2$$

The square has an area of 36 square feet.

In other words, if the square were to be covered in standard size (1 ft^2) floor tiles, the project would require exactly 36 tiles.

Test your skills by solving Example 9. Draw a diagram and label the diagram with the given information. The diagram should help you visualize which formula is appropriate for this problem and identify which variable is unknown.

Example 9: A rancher wants to construct a rectangular pen. He has a 12 foot gate and 148 feet of fencing material. He wants the pen to be 30 feet wide. How long can the pen be?

1. **Draw a picture.**

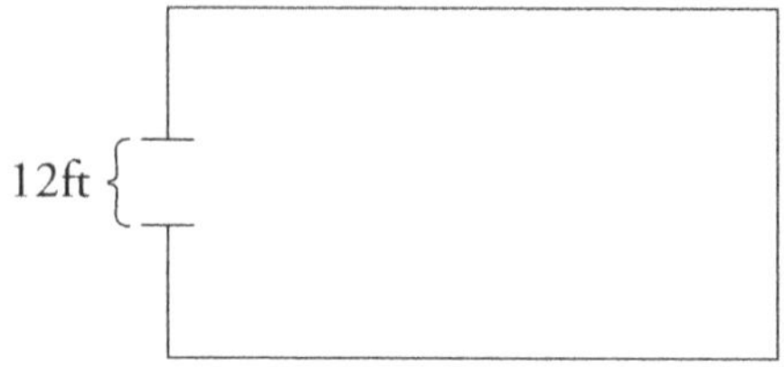

2. **Choose the applicable formula.**

   Which formula applies to this problem? The problem states that the pen is rectangular. Is this an area problem or a perimeter problem? The picture helps determine that this is a perimeter problem. The fencing will be placed around the edge of the rectangle. The appropriate formula for this problem is:

$$P = 2l + 2w$$

3. **Substitute the given information into the formula.**

   The question indicates that the length is the unknown. Numbers are needed for perimeter and width. The problem has three numbers. Does that mean that one of the numbers is extraneous information? No, the distance around the pen will be fencing material and the gate.

$$P = 148 + 12$$
$$P = 160 \text{ ft}$$

4. **Copy the formula for perimeter of a rectangle.**

$$P = 2l + 2w$$

5. **Substitute into the perimeter formula.**

$$160 = 2l + 2(30)$$

6. **Solve the equation one step at a time.**

$$160 = 2l + 60$$
$$160 - 60 = 2l + 60 - 60$$
$$100 = 2l$$
$$\frac{100}{2} = \frac{2l}{2}$$
$$50 = l$$
$$l = 50 \text{ ft}$$

The pen can be no longer than 50 feet.

Multiple Shapes

Some word problems combine more than one shape. These problems are much easier to understand when you use the following basic steps for solving word problems:

1. **Read the problem carefully, looking for keywords.**
2. **Draw a diagram, if possible.**
3. **Label the diagram.**
4. **Write an equation.**
5. **Solve the equation.**
6. **Check your solution in the equation and in the original word problem.**

Follow these steps for Example 10.

Example 10: The length of the side of a square is equal to twice the length of the side of an equilateral triangle. If the perimeter of the square is 32 inches, find the length of the side of the equilateral triangle.

1. **Read the problem carefully.**
2. **Draw pictures of a square and an equilateral triangle.**

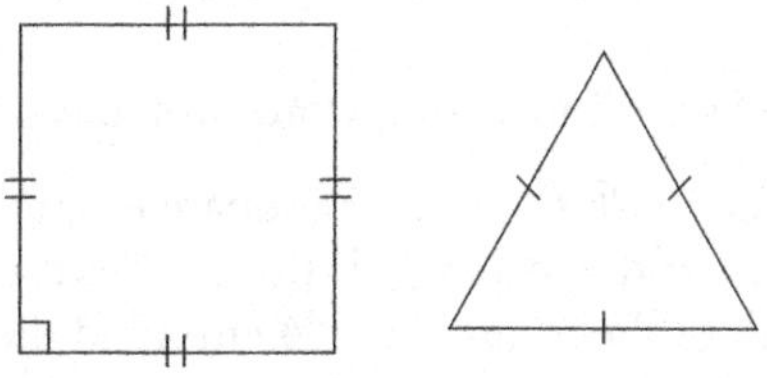

3. **Label the sides of the figures.**

   Because the length of the side of the equilateral triangle is not known, use the variable x to represent the length. The length of the side of the square is defined in relation to the length of the side of the equilateral triangle, x. Often, the unknown at the end of the English sentence is the variable used in the equation.

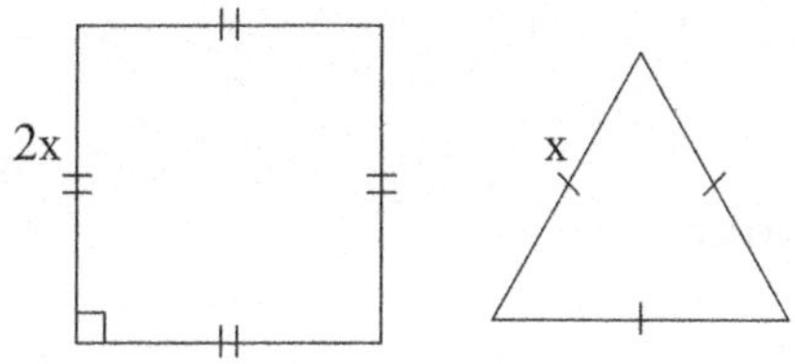

4. **Write the equation, and then substitute into the perimeter formula. The formula for perimeter of a square is**

$$P = 4s$$
$$32 = 4(2x)$$

5. **Solve the equation one step at a time.**

$$32 = 8x$$
$$\frac{32}{8} = \frac{8x}{8}$$
$$4 = x$$
$$x = 4 \text{ in}$$

The length of the side of the equilateral triangle is 4 inches.

6. **Check the solution.**

Check your answer by replacing "the length of the side of an equilateral triangle" with 4 inches and reread the original word problem (see Chapter 4 for details). The problem says that, "the length of the side of a square is equal to twice" 4 inches; therefore, the length of the side of the square is 8 inches. The problem reads, "the perimeter of the square is 32," so the perimeter can be calculated as 4 times 8 inches, which is 32 inches. Therefore, the algebra of solving the equation and the translation of the problem check.

Practice the six basic steps for solving word problems with Example 11.

Example 11: The length of the side of a square is one-half the width of a rectangle. The length of the rectangle is three times the width of the rectangle. The perimeter of the rectangle is 64 cm. Find the length of the side of the square.

1. **Read the problem.**
2. **Draw a picture.**

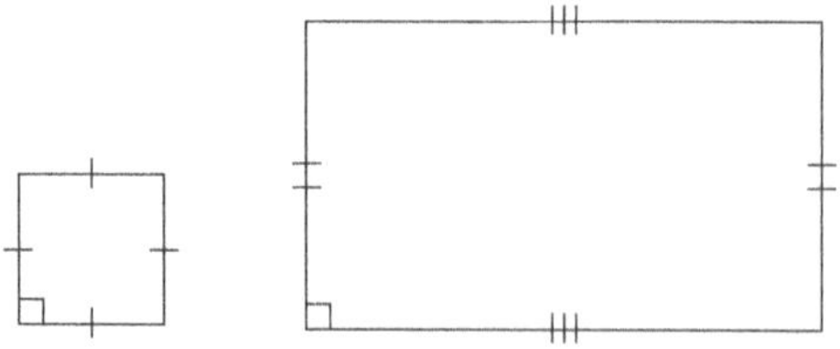

3. **Label the picture.**

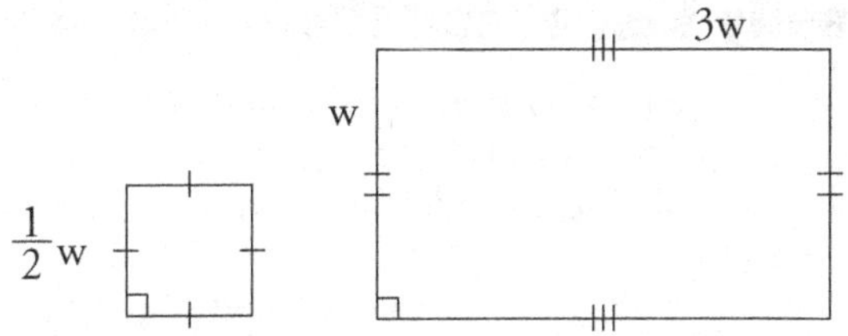

Note: The width of the rectangle is at the end of the first two sentences. The variable w is easier to use than x for this problem because the width (w) is unknown.

4. **Write an equation by choosing the correct formula and substitute information from the picture into the formula.**

The only number given in this problem is the perimeter of a rectangle.

$$P = 2l + 2w$$
$$64 = 2(3w) + 2(w)$$

5. **Solve for the unknown one step at a time.**

$$64 = 6w + 2w$$
$$64 = 8w$$
$$\frac{64}{8} = \frac{8w}{8}$$
$$8 = w$$
$$w = 8 \text{ cm}$$

Is this the correct answer? No! Remember to ask yourself two questions as you complete a word problem.

- Did I answer the question?
- Did I put units on my answer?

The question asked for the length of the side of the square. As you reread the word problem, the diagram you have drawn helps you arrive at the correct answer:

The length of the side of the square is 4 centimeters.

6. **Check the solution.**

Look back at the picture and do mental math to check the dimensions of the rectangle. Does the perimeter add up to 64 centimeters? Check your solution using the picture and always be sure to reread the problem.

Other Easily Visualized Word Problems

Geometry problems require a diagram because they are always concerning a geometric shape. Visualizing a story problem with a diagram is an excellent tool for translating a word problem into the correct equation, even for problems not involving geometry. Board problems may not at first appear to involve a diagram, but drawing and labeling a board makes some word problems easier to solve. See Chapter 7 for details on board problems.

Use the six basic steps for solving word problems (discussed in the "Multiple Shapes" section) for Example 12.

Example 12: Crystal has a 96-inch long board. The directions for a post for her new mailbox require three pieces of wood. The longest piece must be three times the length of the middle-size piece, and the shortest piece must be 9 inches shorter than the middle-size piece. How long must each piece be?

1. **Read the problem.**
2. **Draw a diagram.**

   As you are reading each line of the word problem, add your translation to the diagram.

   First line: Draw a board that is 96 inches long.

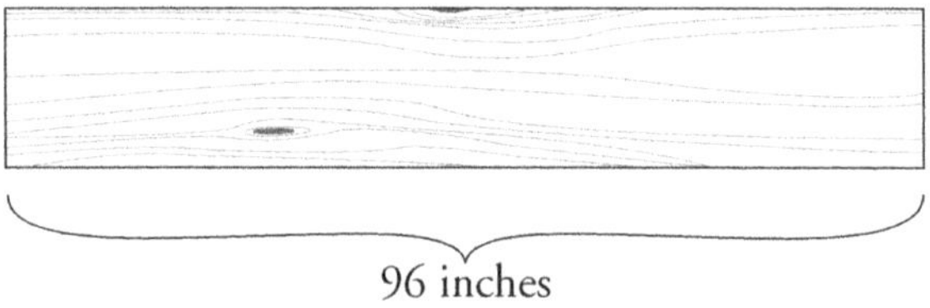

3. **Label the diagram.**

   Second line: Cut the board into three pieces and label the shortest piece, middle-size piece, and the longest piece.

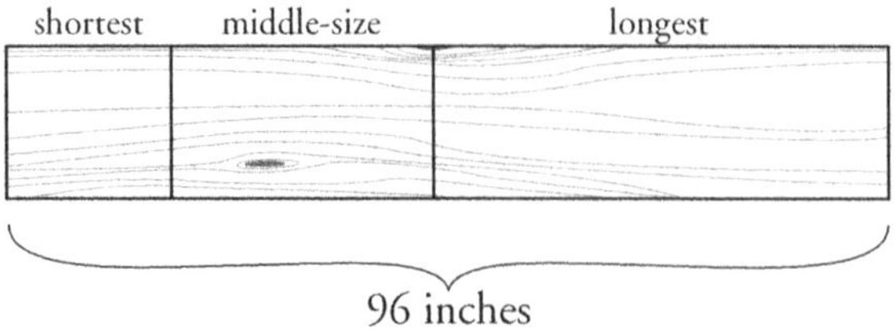

Third line: Put the variables on the board. The first half of the third line tells you the relationship between the longest piece and the middle-size piece. The longest piece is written in relation to the middle-size piece or in terms of the middle-size piece.

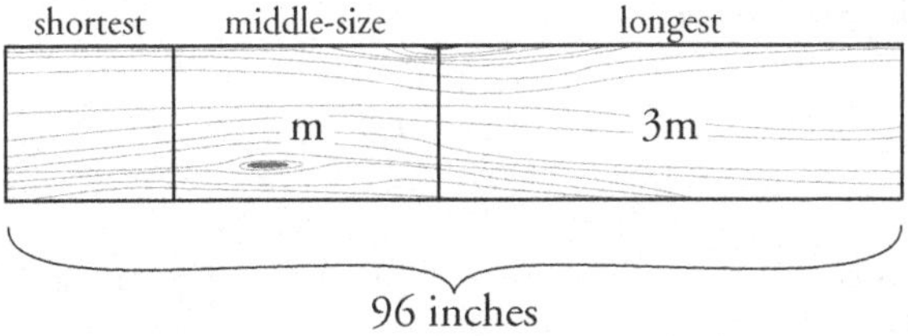

The second half of the third line tells you the length of the shortest piece in terms of the middle-size piece. Don't forget the turnaround word.

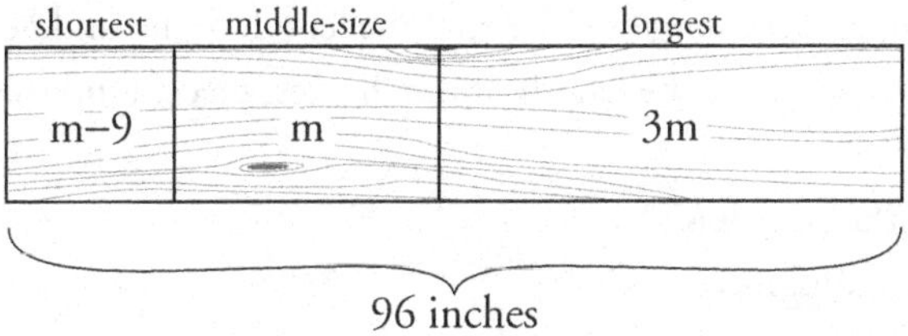

Note: The "middle-size piece" is at the end of the expression and is used as the variable. Also, note that the variable *m* is used instead of an *x* or an *n*. The *m* helps identify the answer to the equation as the middle-size piece. While there are three unknowns, the relationship of two of the unknowns to the middle-size piece is given, so *m* is the only variable in the equation.

4. **Write the equation from the diagram.**

$$(m - 9) + m + 3m = 96$$

5. **Solve the equation one step at a time.**

$$5m - 9 = 96$$
$$5m - 9 + 9 = 96 + 9$$
$$5m = 105$$
$$\frac{5m}{5} = \frac{105}{5}$$
$$m = 21$$

The middle-size piece is 21 inches long. (The lengths of the shortest and longest pieces are then easily found by looking at the diagram and doing mental math.) The shortest piece is 12 inches long. The longest piece is 63 inches long.

6. **Check your answer.**

$$21 + 12 + 63 = 96$$
$$33 + 63 = 96$$
$$96 = 96 ✓$$

Even though Example 13 is not about a board, drawing a picture of a board helps translate the problem correctly.

Example 13: A string knotting project for Troop 47 requires three pieces of string. Each girl has a piece of string that is 42 inches long. The string must be cut into three pieces such that the length of the middle-size piece is $\frac{1}{2}$ the length of the longest piece, and the length of the middle-size piece is 6 inches longer than the length of the shortest piece. Find the length of all three pieces.

1. **Read the problem.**
2. **Draw a diagram.**

   Instead of drawing a string, draw a board.

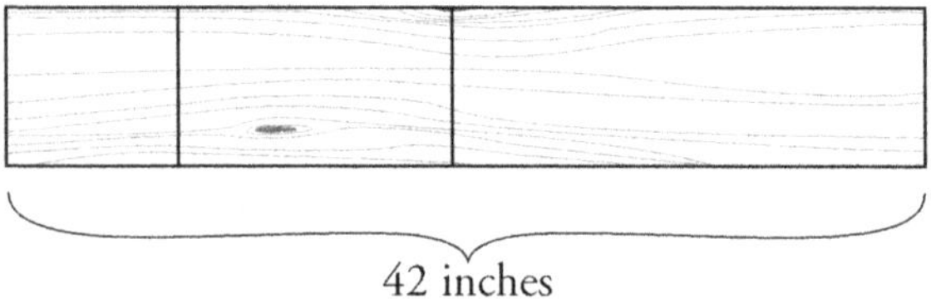

3. **Label each piece with the variable or expression describing the length of that piece in terms of one of the other pieces.**

   The third line defines the relationship of the sizes of the pieces. The first half of line three ends with, "the length of the longest piece." The second half of the third line gives you the length of the middle-size piece in terms of the shortest piece. When all variables are not given in terms of just one of the pieces, the shortest piece is the easiest one to use for the unknown.

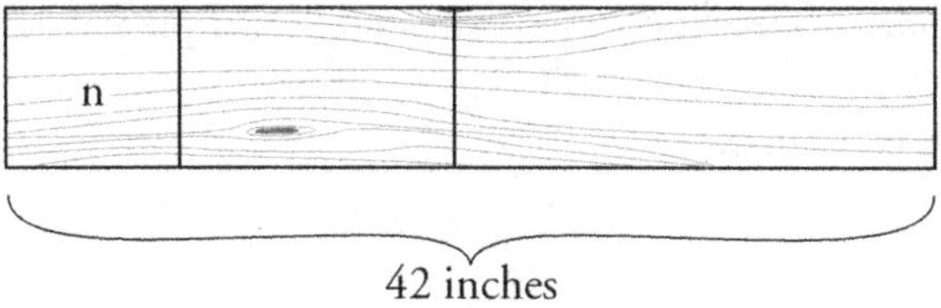

For this problem, use the variable *n* to represent the shortest piece. The variable *s* may seem logical to use for the unknown, but is ill-advised because an *s* can resemble the number 5.

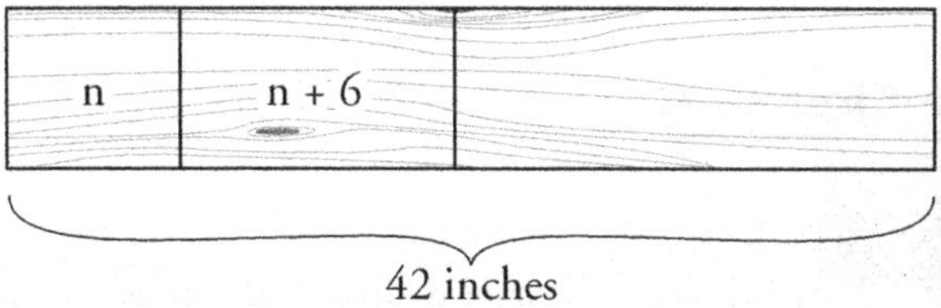

Only the longest piece of the board is left to label. If the middle-size piece is one-half of the length of the longest piece, the longest piece is twice a long as the middle-size piece.

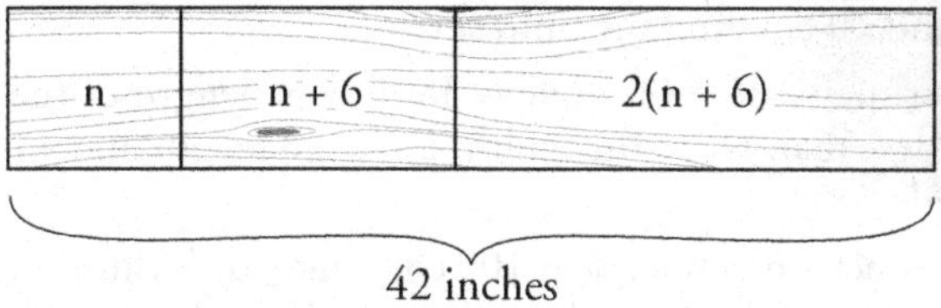

4. **Write the equation.**

$$n + (n + 6) + 2(n + 6) = 42$$

5. **Solve the equation.**

$$n + n + 6 + 2(n + 6) = 42$$

$$n + n + 6 + 2n + 12 = 42$$
$$4n + 18 = 42$$
$$4n + 18 - 18 = 42 - 18$$
$$4n = 24$$
$$\frac{4n}{4} = \frac{24}{4}$$
$$n = 6$$

The shortest piece is 6 inches. The middle-size piece is 12 inches. The longest piece is 24 inches.

6. **Check your answer.**

$$6 + 12 + 24 = 42$$
$$18 + 24 = 42$$
$$42 = 42 \checkmark$$

The solution to the equation has checked. If you want to double-check your answer, reread the problem using 6, 12, and 24 inches to check the translation of the word problem.

Chapter Checkout

Q&A

Solve the following problems. Be sure to answer the question and include units.

1. Find the area of a triangle with a base of 6 m and a height of 10 m.
2. Given the perimeter of a triangle that is 25 cm and two sides measure 7 cm and 10 cm, find the third side.
3. The perimeter of a rectangle is 38 in. The length of the rectangle is 2 in less than twice the width. Find the length and width of the rectangle.
4. The perimeter of a triangle is 100 cm. One side is three times as long as the second side. The third side is six times the second side. Find the measurements of each side.
5. The perimeter of a square is 40 m. Find the length of the side *s*.
6. The area of a rectangle is 50 m^2 and the length is 10 m. Find the width.
7. Jacob mows a yard that is shaped like a rectangle plus a square (see the following figure). The side of the square is $\frac{1}{2}$ the length of the rectangle. The width is two feet less than the length. The entire perimeter of the yard is 126 ft. Find the length of side of the square.

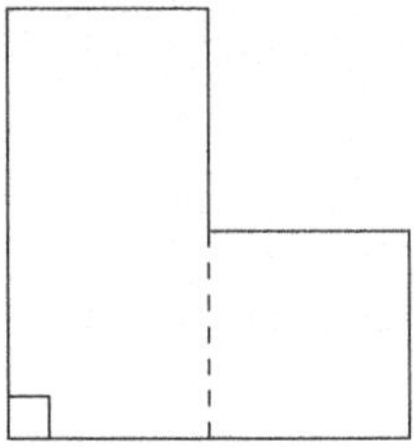

8. The length of the side of a square is twice the width of a rectangle. The length of the rectangle is three times the width of the rectangle. The perimeter of the rectangle is 48 yd. Find the length of the side of the square.

9. Jenna had a bar of candy measuring 8 inches long that she was going to divide into three pieces. She kept the longest piece that was twice as long as the other two pieces of equal size. How long was the piece of candy that Jenna kept?

10. A lumber mill cut trees into four pieces of equal length. If a tree measured 20 ft long, how long was each piece?

11. Michelle wants to decorate the dance hall with streamers for her quinseñera. Michelle wants to use all 250 feet of the streamer. The head table is at one end of the dance hall. She wants two identical streamers across the width of the dance hall meeting above the head table. The other two identical streamers also start above the head table and go to the opposite corners of the dance hall. The streamers going the length of the dance hall need to be four times the length of the shorter streamers. Find the length of all four streamers.

Answers: 1. The area of the triangle is $A = 30\ m^2$ **2.** The third side measures 8 cm. **3.** The width and length of the rectangle are $w = 7$ in, $l = 12$ in **4.** The measurements of the three sides of the triangle are 10 cm, 30 cm, 60 cm. **5.** The length of the side of the square is 10 m. **6.** The width of the rectangle is 5 m. **7.** The length of the square part of the yard is 13 ft. **8.** The length of the side of the square is 12 yd. **9.** Jenna's piece of the candy bar is 4 in long. **10.** Each piece of the tree is 5 ft long. **11.** Two streamers are 25 ft long and the other two streamers are 100 ft long.

Chapter 6
PROPORTIONS AND PERCENTS

Chapter Check-In

- ❑ Proportion problems
- ❑ Proportion word problems
- ❑ Percent problems
- ❑ Percent word problems

Word problems involving **proportions** or **percentages** are some of the easiest to translate into equations and solve.

In this chapter, you solve equations involving proportions and **percents,** and you also recognize and solve word problems that contain proportions and percents.

Proportion Problems

Many standardized tests contain a section on **analogies,** in which you are asked to fill in the blank. An example of this type of problem is

sky is to blue as grass is to ________

You then fill in the blank with "green" because the relationship between sky and blue is the same as the relationship between grass and green.

In math, a similar example may be something like this

one is to three as two is to six

The relationship between one and three is that the second number is three times as large as the first number. In math, we call that relationship a **ratio.** The second ratio, two is to six, has the same relationship as the first ratio. When two ratios are equal, that is called a **proportion.** Rather than write a proportion in English, you can write the proportion "one is to three as two is to six" as follows: $\frac{1}{3} = \frac{2}{6}$. Notice that if the second fraction is reduced, both fractions are $\frac{1}{3}$.

When solving proportion problems, one of the values is unknown and is, therefore, assigned a **variable,** such as *x* or *p*. To solve for that variable, you must solve an equation containing fractions. When an equation has two fractions set equal to each other, you can simplify both sides of the equation by **clearing fractions.** To clear fractions from an equation, find the **least common denominator** (LCD) for all the fractions. Multiply both sides of the equation by the LCD.

Example 1 illustrates the steps required to solve a proportion problem.

Example 1: Solve for *x*: $\frac{3}{4} = \frac{x}{20}$

1. **Simplify the equation by clearing fractions.**

   The two denominators in this problem are 4 and 20. The **least common multiple** of 4 and 20 is 20: 4 times 5 is equal to 20, and 20 times 1 is equal to 20. Writing the least common denominator over 1 makes canceling easier.

$$\frac{20}{1} \times \frac{3}{4} = \frac{20}{1} \times \frac{x}{20}$$

$$\frac{\overset{5}{\cancel{20}}}{1} \times \frac{3}{\underset{1}{\cancel{4}}} = \frac{\overset{1}{\cancel{20}}}{1} \times \frac{x}{\underset{1}{\cancel{20}}}$$

$$15 = 1x$$

2. **Divide each side of the equation by the coefficient of x.**

$$\frac{1x}{1} = \frac{15}{1}$$

$$x = 15$$

You can check a proportion problem by substituting the solution into the original equation and reducing all fractions. The following shows how to check the solution:

$$\frac{3}{4} = \frac{15}{20}$$

$$\frac{3}{4} = \frac{3}{4} \checkmark$$

Practice your proportion solving skills with Example 2.

Example 2: Solve the following proportion for *x* and check your answer:

$$\frac{x}{27} = \frac{2}{9}$$

1. **Simplify the equation by clearing fractions.**

   The least common denominator of 27 and 9 is 27.

$$\frac{27}{1} \times \frac{x}{27} = \frac{27}{1} \times \frac{2}{9}$$

$$\frac{\overset{1}{\cancel{27}}}{1} \times \frac{x}{\underset{1}{\cancel{27}}} = \frac{\overset{3}{\cancel{27}}}{1} \times \frac{2}{\underset{1}{\cancel{9}}}$$

$$1x = 6$$

2. **Divide each side of the equation by the coefficient of x.**

$$\frac{1x}{1} = \frac{6}{1}$$

$$x = 6$$

3. **Check.**

$$\frac{6}{27} = \frac{2}{9}$$

$$\frac{2}{9} = \frac{2}{9} \checkmark$$

Example 3 uses the same steps, but this time with the unknown in the denominator instead of the numerator.

Example 3: Solve and check the following proportion for *x*: $\frac{32}{x} = \frac{8}{5}$

1. **Simplify the equation by clearing fractions.**

   The least common denominator of *x* and 5 is 5*x*. For a review of least common denominators with variables, see *CliffsQuickReview Algebra I,* by Jerry Bobrow (Wiley).

$$\frac{5x}{1} \times \frac{32}{x} = \frac{5x}{1} \times \frac{8}{5}$$

$$\frac{\overset{1}{5\cancel{x}}}{1} \times \frac{32}{\underset{1}{\cancel{x}}} = \frac{\overset{1}{\cancel{5}x}}{1} \times \frac{8}{\underset{1}{\cancel{5}}}$$

$$160 = 8x$$

2. Divide each side of the equation by the coefficient of x.

$$\frac{8x}{8} = \frac{160}{8}$$
$$x = 20$$

3. Check.

$$\frac{32}{20} = \frac{8}{5}$$
$$\frac{8}{5} = \frac{8}{5} \checkmark$$

Proportion Word Problems

When word problems state a relationship between two quantities, they provide you with a ratio. That ratio can be used to set up a proportion and solve for an unknown. After you have identified the ratio, the rest of the equation can be set up using that pattern. Example 4 shows the steps required to set up, solve, and check a proportion word problem.

Example 4: Missy's car averages 32 miles per gallon. She wants to drive home on Christmas day and wants to be sure she has enough gas to get home without having to stop for fuel. Money is tight, and her car is on empty. How many gallons of gas should she buy to be sure she can make the 224-mile trip home?

1. **Identify the ratio given in the word problem.**

   The ratio is 32 miles per gallon. The ratio written as a fraction is

$$\frac{32 \text{ miles}}{1 \text{ gallon}}$$

2. **Write the proportion equation, following the pattern set by the ratio.**

   The ratio has number of miles in the numerator and number of gallons in the denominator. The other information given is the number of miles, so that number is placed in the numerator.

$$\frac{32 \text{ miles}}{1 \text{ gallon}} = \frac{224 \text{ miles}}{x}$$

   Note: After the equation is set up, the units are not needed to find the solution. They are needed, however, when stating the answer.

3. **Simplify the equation by clearing fractions.**

   The least common denominator of 1 and x is $1x$.

$$\frac{1x}{1} \times \frac{32}{1} = \frac{1x}{1} \times \frac{224}{x}$$

$$\frac{\overset{1}{\cancel{1}x}}{1} \times \frac{32}{\underset{1}{\cancel{1}}} = \frac{\overset{1}{1\cancel{x}}}{1} \times \frac{224}{\underset{1}{\cancel{x}}}$$

$$32x = 224$$

4. **Divide each side of the equation by the coefficient of x.**

$$\frac{32x}{32} = \frac{224}{32}$$
$$x = 7$$

Missy must buy at least 7 gallons of gas to be able to get home without stopping for gas.

5. **Check.**

$$\frac{32}{1} = \frac{224}{7}$$
$$\frac{32}{1} = \frac{32}{1} \checkmark$$

You may have been able to work Example 4 without an equation. Some proportion problems are easy enough to work in your head; these are good problems to practice the skills necessary to solve the more difficult proportion word problems.

Use Example 5 as another way to practice your problem-solving skills.

Example 5: The box of just-add-water pancake mix says to mix one cup of water with two cups of pancake mix. Bruce is in charge of the preparation of the pancakes for a pancake supper fund-raiser. Each cook has a six quart container for the batter. How many quarts of water should be added to four quarts of pancake mix?

1. **Identify the ratio given in the word problem.**

The ratio is one cup of water to two cups of pancake mix. The ratio written as a fraction is:

$$\frac{1 \text{ cup water}}{2 \text{ cups pancake mix}}$$

Note: A detailed description of the units must be included in this example, because both water and pancake mix are measured in cups.

2. **Write the proportion equation following the pattern set by the ratio.**

   The ratio has volume of water in the numerator and volume of pancake mix in the denominator. The other information given in the problem is six quart container and four quarts of pancake mix. How do you know which number to put in the equation? Follow the pattern of the ratio. The six quart container holds the batter; it is neither the volume of water nor the volume of pancake mix. The four quarts is pancake mix and, therefore, is placed in the denominator. Notice that you are solving for the volume of water because the unknown is in the numerator.

$$\frac{1 \text{ cup water}}{2 \text{ cups pancake mix}} = \frac{x}{4 \textit{ quarts pancake mix}}$$

   Note: After the equation is set up, the units are not needed to find the solution. They are needed, however, when stating the answer.

3. **Simplify the equation by clearing fractions.**

   The least common denominator of 2 and 4 is 4.

$$\frac{4}{1} \times \frac{1}{2} = \frac{4}{1} \times \frac{x}{4}$$

$$\frac{\overset{2}{\cancel{4}}}{1} \times \frac{1}{\cancel{2}_1} = \frac{\overset{2}{\cancel{4}}}{1} \times \frac{x}{\cancel{4}_1}$$

$$2 = x$$

   Bruce needs to add two quarts of water to every four quarts of pancake mix.

4. **Check.**

$$\frac{1}{2} = \frac{2}{4}$$

$$\frac{1}{2} = \frac{1}{2} \checkmark$$

In Example 6, you work a problem that illustrates the importance of setting up a proportion.

Example 6: The quality-control supervisor at Papel Paper Products has calculated that, out of 5,000 reams of copy paper, 100 reams are usually defective. How many reams of paper does the quality control supervisor believe must be produced to fill an order of 2,450 reams that are not defective?

1. **Identify the ratio given in the word problem.**

   The ratio is 100 defective reams for every 5,000 total reams. The ratio written as a fraction is:

$$\frac{100 \text{ defective reams}}{5{,}000 \text{ total reams}}$$

2. **Write the proportion equation following the pattern set by the ratio.**

   The ratio has number of defective reams in the numerator and number of total reams in the denominator. Can you find another number of defective reams or total number of reams in the word problem? No, 2,450 is the number of nondefective reams in the order. Is there another ratio that can be written from the original problem? Yes, you can calculate the number of nondefective reams out of the 5,000. 4,900 reams out of 5,000 are not defective (5,000 – 100 = 4,900).

$$\frac{4,900 \textit{ nondefective reams}}{5,000 \textit{ total reams}} = \frac{2,450 \textit{ nondefective reams}}{x}$$

   Note: After the equation is set up, the units are not needed to find the solution. They are needed, however, when stating the answer.

3. **Simplify the equation by clearing fractions.**

   The least common denominator of 5,000 and x is 5,000x.

$$\frac{5,000}{1} \times \frac{4,900}{5,000} = \frac{5,000}{1} \times \frac{2,450}{x}$$

$$\frac{\overset{1}{\cancel{5{,}000}}x}{1} \times \frac{4{,}900}{\underset{1}{\cancel{5{,}000}}} = \frac{5{,}000\cancel{x}^{1}}{1} \times \frac{2{,}450}{\underset{1}{\cancel{x}}}$$

$$4{,}900x = 12{,}250{,}000$$

4. **Divide each side of the equation by the coefficient of x.**

$$\frac{4,900x}{4,900} = \frac{12,250,000}{4,900}$$

$$x = 2{,}500$$

   Papel Paper Products must produce over 2,500 reams to ensure that it can fill the order for 2,450 nondefective reams.

5. **Check.**

$$\frac{4,900}{5,000} = \frac{2,450}{2,500}$$

$$\frac{49}{50} = \frac{49}{50} \checkmark$$

Percent Problems

The word **percent** means per one hundred. One method of solving percent problems is the percent proportion method. This section reviews using a proportion to solve percent problems.

The three elements needed in a percent problem are the percent, the **base,** and the **amount.** When one of these is unknown (but no more than one), you can use the percent proportion method to find the missing element.

The first clue that a problem is a percent problem is the word percent or the percent (%) sign. To work a percent problem, use the following steps:

1. **Identify the percent.**

   The percent is the number before the word "percent" or the percent (%) sign.

2. **Identify the base.**

   The base is the number after the word OF.

3. **Identify the amount.**

   The amount is hardest to identify and is one of the following:

   - The last number in the problem
   - The number after the word IS

4. **Enter the values in the percent proportion formula.**

   The percent proportion formula is $\frac{a}{b} = \frac{p}{100}$, where a stands for the amount, b stands for the base, and p is the percent.

5. **Solve the equation for the unknown.**

   The same steps used to solve proportions are used to solve for the unknown value. If the percent is the unknown, be sure to include a % sign with your answer.

Practice the percent proportion method by solving Example 7:

Example 7: What percent of 28 is 7?

1. **Identify the percent.**

   The percent is the unknown in this problem.

   $$p = ?$$

2. **Identify the base.**

   The base is the number after the word OF, 28.

   $$b = 28$$

3. **Identify the amount.**

   The amount is the number after the word IS, 7.

   $$a = 7$$

4. **Enter the values in the percent proportion formula.**

   $$\frac{7}{28} = \frac{p}{100}$$

5. **Solve the equation for the unknown.**

   The fraction $\frac{7}{28}$ can be reduced to $\frac{1}{4}$.

   $$\frac{1}{4} = \frac{p}{100}$$

   The LCD of 4 and 100 is 100.

   $$\frac{100}{1} \times \frac{1}{4} = \frac{100}{1} \times \frac{p}{100}$$

   $$\frac{\overset{25}{\cancel{100}}}{1} \times \frac{1}{\underset{1}{\cancel{4}}} = \frac{\overset{1}{\cancel{100}}}{1} \times \frac{p}{\underset{1}{\cancel{100}}}$$

   $$25 = p$$

   $$p = 25\%$$

   25% of 28 is 7.

The same steps can be used even if the percent is given and another element of the percent proportion is missing, as you see in Example 8:

Example 8: What is 45% of 20?

1. **Identify the percent.**

   The percent is 45.

   $$p = 45$$

2. **Identify the base.**

   The base is the number after the word OF, 20.

   $$b = 20$$

3. **Identify the amount.**

   The amount is the unknown.

   $$a = ?$$

4. **Enter the values in the percent proportion formula.**

$$\frac{a}{20} = \frac{45}{100}$$

5. **Solve the equation for the unknown.**

The LCD of 20 and 100 is 100.

$$\frac{100}{1} \times \frac{a}{20} = \frac{100}{1} \times \frac{45}{100}$$

$$\frac{\overset{5}{\cancel{100}}}{1} \times \frac{a}{\underset{1}{\cancel{20}}} = \frac{\overset{1}{\cancel{100}}}{1} \times \frac{45}{\underset{1}{\cancel{100}}}$$

$$5a = 45$$

$$\frac{5a}{5} = \frac{45}{5}$$

$$a = 9$$

9 is 45% of 20.

In Example 9, you solve for a different unknown.

Example 9: 96 is 120% of what number?

1. **Identify the percent.**

$$p = 120$$

2. **Identify the base.**

The words "what number" come after OF; therefore, the base is the unknown.

$$b = ?$$

3. **Identify the amount.**

The amount, 96, is the last number in the problem.

$$a = 96$$

4. **Enter the values in the percent proportion formula.**

$$\frac{96}{b} = \frac{120}{100}$$

5. **Solve the equation for the unknown.**

The LCD of b and 100 is $100b$.

$$\frac{100b}{1} \times \frac{96}{b} = \frac{100b}{1} \times \frac{120}{100}$$

$$\frac{100\cancel{b}^{1}}{1} \times \frac{96}{\cancel{b}_{1}} = \frac{\cancel{100b}^{1}}{1} \times \frac{120}{\cancel{100}_{1}}$$

$$9{,}600 = 120b$$

$$\frac{9,600}{120} = \frac{120b}{100}$$

$$b = 80$$

96 is 120% of 80.

Note: The amount is always larger than the base when the percent is more than 100.

Percent Word Problems

To solve word problems that involve percentages, use the same steps as in the preceding section:

1. **Identify the percent.**
2. **Identify the base.**
3. **Identify the amount.**
4. **Enter the values in the percent proportion formula.**
5. **Solve the equation for the unknown.**

Practice these five steps by solving Example 10.

Example 10: Kay is in the 20% income tax bracket. Last year, her taxable income was $64,800. How much will she need to pay in taxes?

1. **Identify the percent.**

   The percent is 20%.

$$p = 20$$

2. **Identify the base.**

   Even though the base is not explicitly stated, you can think of the problem as if it said, "Kay's income tax is 20% of her taxable income." Because the words "taxable income" follows OF, that is the base.

$$b = 64{,}800$$

3. **Identify the amount.**

   The amount is the unknown.

$$a = ?$$

4. **Enter the values in the percent proportion formula.**

$$\frac{a}{64,800} = \frac{20}{120}$$

5. **Solve the equation for the unknown.**

   The LCD of 64,800 and 100 is 64,800.

$$\frac{64,800}{1} \times \frac{a}{64,800} = \frac{64,800}{1} \times \frac{20}{100}$$

$$\frac{\overset{1}{\cancel{64,800}}}{1} \times \frac{a}{\underset{1}{\cancel{64,800}}} = \frac{\overset{648}{\cancel{64,800}}}{1} \times \frac{20}{\underset{1}{\cancel{100}}}$$

$$a = 648 \times 20$$
$$a = 12,960$$

   Kay will need to pay $12,960 in taxes.

To help you identify the parts of the percent proportion, use one ratio as a pattern for the ratio on the other side of the equation. Example 11 shows how to use one ratio as a pattern.

Example 11: In the local election, 4,775 voters cast their votes in the election for mayor. 52% of the voters voted for incumbent Mayor Harris. How many did not vote for Mayor Harris?

1. **Identify the percent.**

   The percent voting for Harris was 52%.

$$p = 52$$

2. **Identify the base.**

   Notice the word problem says 52% of the voters. Because "the voters" follows OF, that is the base.

$$b = 4,775$$

3. **Identify the amount.**

   The amount is the unknown.

$$a = ?$$

4. **Enter the values in the percent proportion formula.**

$$\frac{a}{4,775} = \frac{52}{100}$$

The ratio $\frac{52}{100}$ gives the percentage of voters for Harris in the numerator and the total percentage of voters in the denominator. The other ratio must follow the same pattern: the number of voters for Harris over the total number of voters. You know the total number of voters and can solve for the number of voters *for* Harris. Is that what the question asked? No, it asked for the number of voters who *did not* vote for Harris.

Can you find the percentage of voters who did not vote for Harris? Yes: 48% did not vote for Harris. (100% – 52% = 48%). **Note:** 48% is the **complement** of 52%.

To find the number of voters who did not vote for Harris, solve this percent proportion:

$$\frac{a}{4,775} \times \frac{48}{100}$$

5. **Solve the equation for the unknown.**

A common denominator of 4,775 and 100 is 477,500 (the product of 4,775 and 100).

$$\frac{477,500}{1} \times \frac{a}{4,775} = \frac{477,500}{1} \times \frac{48}{100}$$

$$\frac{\overset{100}{\cancel{477,500}}}{1} \times \frac{a}{\underset{1}{\cancel{4,775}}} = \frac{\overset{4775}{\cancel{477,500}}}{1} \times \frac{48}{\underset{1}{\cancel{100}}}$$

$$100a = 4,775 \times 48$$

$$100a = 229,200$$

$$\frac{100a}{100} = \frac{229,200}{100}$$

$$a = 2,292$$

2,292 voters did not vote for Harris.

If you had solved the original proportion, $\frac{a}{4,775} = \frac{52}{100}$, you would have found that $a = 2,483$. Then, ask yourself two questions:

- Did I answer the question?
- Did I put units on my answer?

Asking those questions makes you realize that 2,483 voters cast their votes *in favor* of Mayor Harris. The question asked you to find the number of voters who did *not* vote for Harris. You can still find that number by subtracting the number who voted for Harris from the total.

$$4{,}775 - 2{,}483 = 2{,}292$$

2,292 voters did not vote for Harris.

Chapter Checkout

Q&A

Solve the following problems for the unknown.

1. $\frac{4}{21} = \frac{x}{63}$
2. $\frac{y}{8} = \frac{5}{2}$
3. $\frac{10}{x} = \frac{4}{10}$
4. The Happy Tree Day Care wants to have 1 teacher for every 6 toddlers. If it has an enrollment of 48 toddlers, how many teachers should be working each day?
5. For every 3 sports cars sold at the local car dealership, 6 SUVs are sold. If the dealership predicts that it will sell 8 sports cars next week, how many SUVs will it probably sell next week?
6. Mac's Mattress Co. produces twin size mattresses. The company is proud to say that only 2 mattresses out of 800 produced are flawed and cannot be sold. In order to fill an order of 399 mattresses for a new dorm, how many mattresses should the company make?
7. 12 is what percent of 60?
8. What is 110% of 40?
9. 9 is 25% of what number?
10. The Creamy Ice Cream Factory is known for its homemade vanilla ice cream. Even though it makes many wonderful flavors of ice cream, 54% of all ice cream made at the factory is homemade vanilla. If 4,050 gallons of homemade vanilla were made last month, what was the total number of gallons of all flavors of ice cream made that month?
11. Donald is a real estate agent who receives 6% commission on each sale he makes. If he sells a ranch for $458,000, how much commission will he receive for the sale?

Answers: **1.** $x = 12$ **2.** $y = 20$ **3.** $x = 25$ **4.** The Happy Tree Day Care should have 8 teachers working each day. **5.** The dealership should expect to sell 16 SUVs next week. **6.** The mattress factory should make at least 400 mattresses to make sure they can fill the order. **7.** 12 is 20% of 60. **8.** 44 is 110% of 40. **9.** 9 is 25% of 36.**10.** The Creamy Ice Cream Factory made 7,500 total gallons of ice cream that month. **11.** Donald will receive $27,480 in commission for the sale of the ranch.

Chapter 7

SUMMATION PROBLEMS USING THE BOARD METHOD

Chapter Check-In

- ❑ Simple board problems
- ❑ Consecutive integers
- ❑ Consecutive even and odd integers
- ❑ Other problems that can be represented with a board

Most problem solvers have more difficulty arriving at a correct equation than solving an equation itself. Drawing a diagram of the word problem helps make the translation easier. In this chapter, you find a variety of word problems that can be represented with a diagram called a board.

When a problem has a total — even if the word "total" is not used — a board can be used to visualize the translation. Summation problems, problems with a total, can be easily represented with a diagram. Hence the name of this chapter. Follow these basic steps (discussed in Chapter 5) for solving word problems — note that Steps 2 and 3 involve drawing and labeling a diagram:

1. **Read the problem carefully, looking for keywords.**
2. **Draw a diagram, if possible.**
3. **Label the diagram.**
4. **Write an equation.**
5. **Solve the equation.**
6. **Check your solution in the equation and in the original word problem.**

Simple Board Problems

Because the diagrams associated with these problems often look like a piece of wood — a board — these problems are known as board problems.

Example 1: Park Ranger Ricki is making signs for the trails at Oak Ridge State Park. She needs to make a sign from a 70-inch treated 2 x 4. The sign will have two posts of equal length, and the sign itself will be $\frac{3}{2}$ of the length of a post. Find the length of each post and of the sign.

1. **Read the problem, looking for keywords.**
2. **Draw a diagram, if possible.**

   Draw a 70-inch-long board.

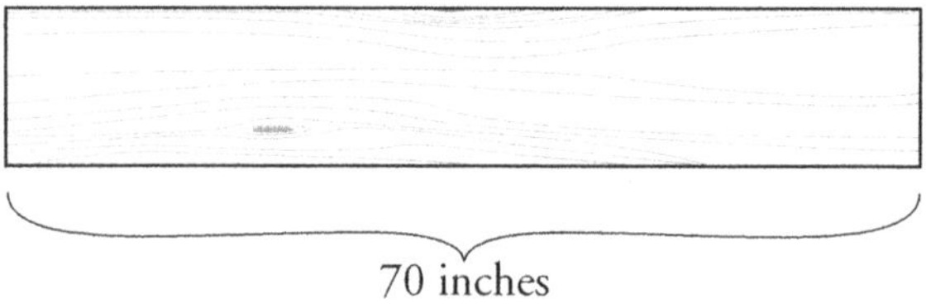

   Cut the board into three pieces.

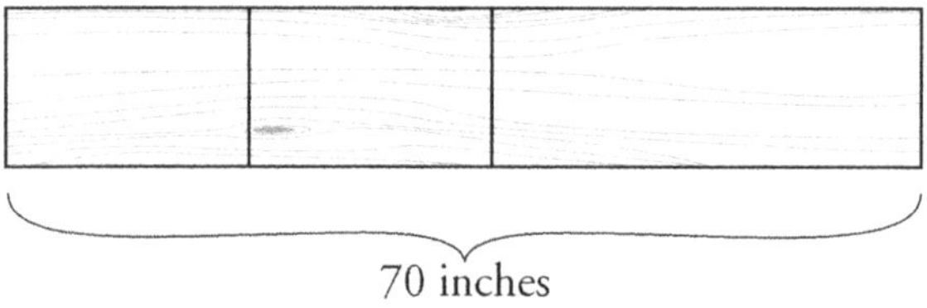

3. **Label the diagram.**

   Use the variable p (for post) to represent the length of the post. The length of the post is chosen as the unknown because the length of the sign is given in terms of the post and the problem says "length of the post" at the end of the sentence.

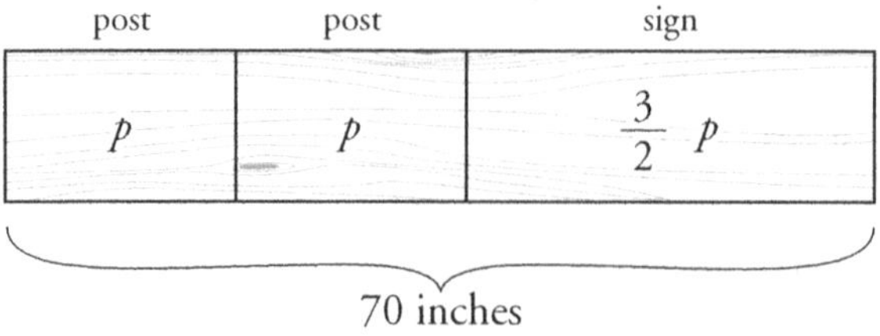

   Note: The keyword OF follows a fraction and, therefore, indicates multiplication.

4. Write an equation.

$$p+p+\frac{3}{2}p=70$$

5. Solve the equation.

The steps for solving a linear equation (discussed in Chapter 4) are as follows:

a. Simplify both sides of the equation by using the distributive property and combining like terms, if possible.

When an equation has a fraction, simplify both sides of the equation by using the distributive property, **clearing fractions,** and combining like terms. To clear fractions from an equation, find the **least common denominator** for all the fractions. Multiply both sides of the equation by the least common denominator.

b. Move all the terms with variables to one side of the equation using the addition property of equations.

c. Move the constants to the other side of the equation using the addition property of equations.

d. Divide by the coefficient using the multiplication property of equations.

e. Check your solution.

The equation has only one fraction, and that fraction has a denominator of 2. Multiply both sides of the equation by 2 and distribute the 2 to every term on the left side of the equation.

$$2\left(p+p+\frac{3}{2}p\right)=2(70)$$

$$2p+2p+2\left(\frac{3}{2}\right)p=140$$

$$2p+2p+3p=140$$

$$7p=140$$

$$\frac{7p}{7}=\frac{140}{7}$$

$$p=20$$

After you have an answer, ask yourself:

Did I answer the question?

Did I include the units?

Each post will be 20 inches tall. To find the length of the sign, look at the diagram of the board in Step 2. The sign is $\frac{3}{2}$ of the length of the post.

$$\frac{3}{2} \times 20$$

$$\frac{3}{\cancel{2}_1} \times \frac{\cancel{20}^{10}}{1}$$

$$30$$

Answer the question and include the units: The length of the sign is 30 inches; each post is 20 inches long.

6. **Check the solution.**

   a. Check the solution in the equation:

$$20 + 20 + \frac{3}{2}(20) = 70$$
$$20 + 20 + 30 = 70$$
$$40 + 30 = 70$$
$$70 = 70 \checkmark$$

   b. Check the solution in the original word problem and with the picture:

   Park Ranger Ricki is making signs for the trails at Oak Ridge State Park. She needs to make a sign from a 70 inch treated 2 x 4. The sign will have two posts of *20 inches*, and the sign itself will be $\frac{3}{2}$ of the length of a post, *30 inches*. Find the length of each post and the sign.

   Mental math reveals that 20 + 20 + 30 is a 70-inch-long board.

Try Example 2 on your own first, and then check yourself with the steps that follow the example.

Example 2: Bob wants to build a bookcase from a 12-foot 1 x 10. He has drawn a rough sketch of the bookshelf. The top of the bookcase should be 2 inches longer than the three shelves. The sides of the bookcase need to be exactly 39 inches long. How long can the shelves and top be?

1. **Read the problem, looking for keywords.**
2. **Draw a diagram, if possible.**

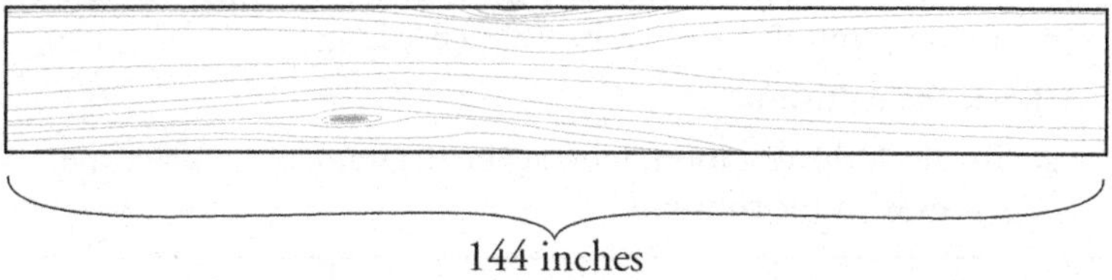

All the lengths must be measured using the same units. Convert 12 feet to 144 inches by multiplying 12 ft by 12 $\frac{\text{in}}{\text{ft}}$, because there are 12 inches in a foot.

The rough sketch shows that the bookshelf will take six pieces of 1 x 10 board.

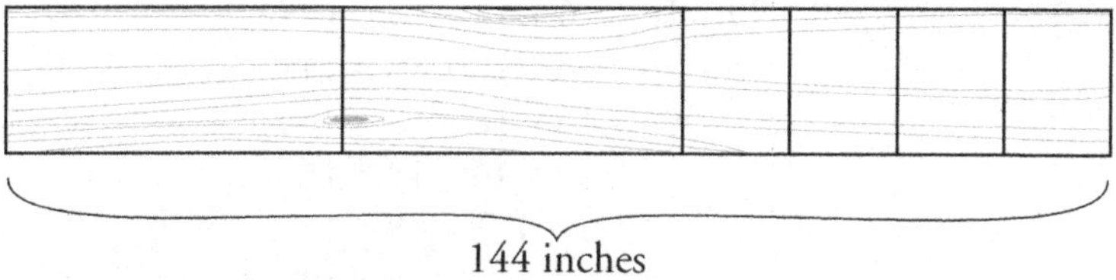

3. **Label the diagram.**

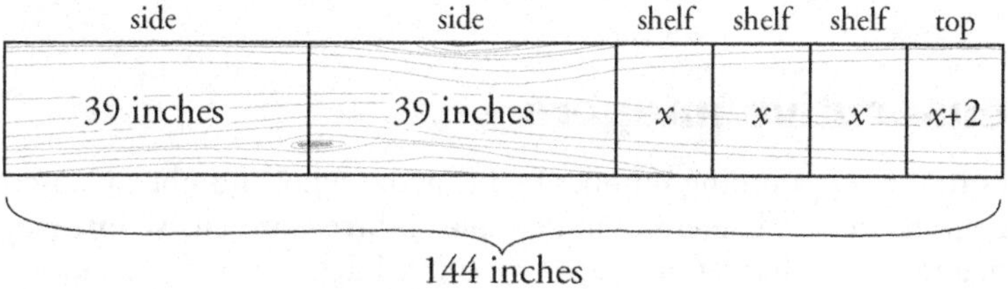

The length of the sides is a constant 39 inches. The length of the shelf is chosen to be the unknown variable x. The length of the top is given in terms of the length of the shelves.

4. **Write the equation.**

$$39 + 39 + x + x + x + (x + 2) = 144$$

5. **Solve the equation.**

$$\begin{aligned} 78 + 4x + 2 &= 144 \\ 4x + 80 &= 144 \\ 4x + 80 - 80 &= 144 - 80 \\ 4x &= 64 \\ \frac{4x}{4} &= \frac{64}{4} \\ x &= 16 \end{aligned}$$

Answer the question and include the units: Each shelf can be 16 inches long, and the top can be 18 inches long.

6. **Check the solution.**

   a. To check the equation, look at the diagram and replace the pieces with the solution:

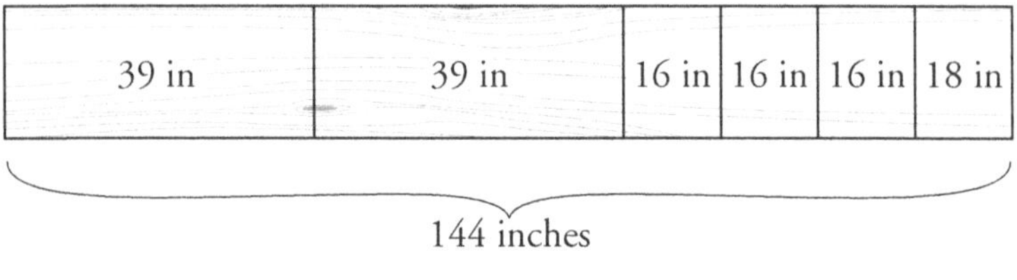

$$\begin{aligned} 39 + 39 + 16 + 16 + 16 + 18 &= 144 \\ 78 + 16 + 16 + 16 + 18 &= 144 \\ 94 + 16 + 16 + 18 &= 144 \\ 110 + 16 + 18 &= 144 \\ 126 + 18 &= 144 \\ 144 &= 144 \checkmark \end{aligned}$$

   b. To check the translation, reread the problem.

Consecutive Integers

An **integer** is a counting number or a negative whole number. Consecutive means "next." Therefore, an example of three **consecutive integers** is 4, 5, and 6. Another set of three consecutive integers is –7, –6, and –5.

If the first integer is unknown, use the variable x. The second integer is one more than the first, or $x + 1$. The third integer is then $x + 2$. Notice this relationship is true whether you consider the set of three positive consecutive integers or the set of three negative consecutive integers. If x is 4, the second integer is $x + 1$, or 5. The third integer is $x + 2$, or 6. This is also true for negative numbers. If the first integer, x, is –7, $x + 1$ is –6, and $x + 2$ is –5.

When working consecutive integer word problems, use x, $x + 1$, and $x + 2$ to represent three unknown consecutive integers. If the problem asks only for two consecutive integers, use x and $x + 1$.

Example 3: Find two consecutive integers such that the total of five times the second integer and twice the first integer is nineteen.

1. **Read the problem, looking for keywords.**

   Notice the leading keyword, TOTAL OF. Because there is a total, a board can be used to diagram this word problem.

2. **Draw a diagram, if possible.**

   In this example, the total is 19. Draw a board 19 long.

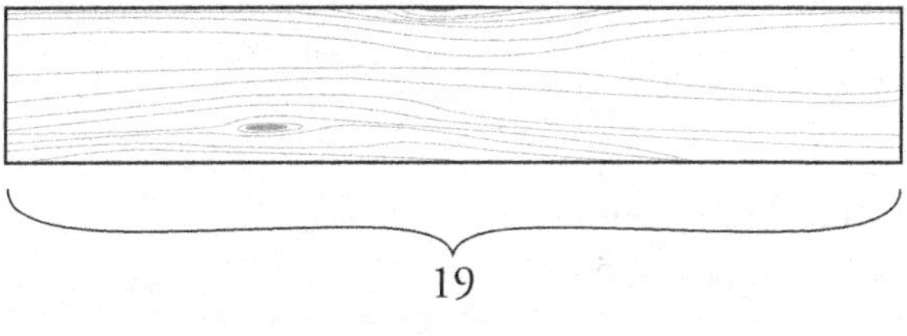

   How many things are added up to get 19? Two.

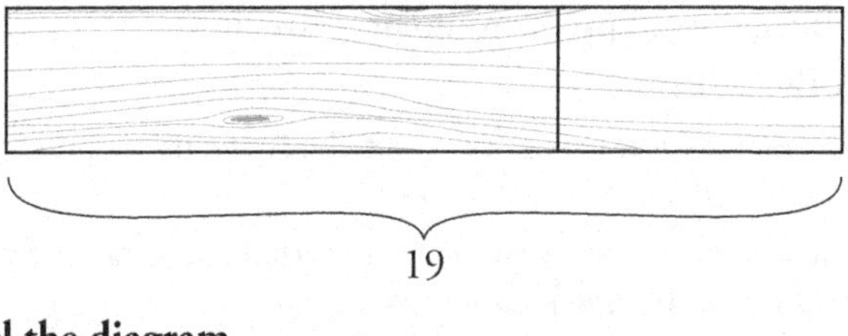

3. **Label the diagram.**

   Choose the first integer to be the unknown, x.

   The second integer is $x + 1$.

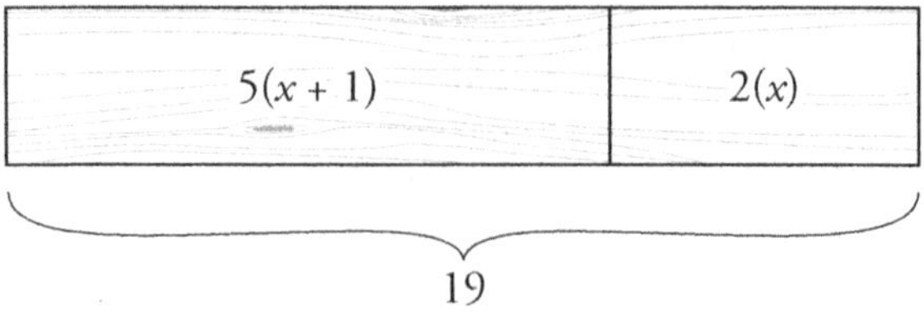

Note: Translation skills developed in Chapter 1 are necessary to label the diagram correctly.

4. **Write the equation.**

$$5(x + 1) + 2(x) = 19$$

5. **Solve the equation.**

$$\begin{aligned} 5(x + 1) + 2x &= 19 \\ 5x + 5 + 2x &= 19 \\ 7x + 5 &= 19 \\ 7x + 5 - 5 &= 19 - 5 \\ 7x &= 14 \\ \frac{7x}{7} &= \frac{14}{7} \\ x &= 2 \end{aligned}$$

Answer the question and include the units:

The first integer is 2.

The second integer is 3.

Note: Integer word problems do not have units.

6. **Check the solution.**

Reread the question using "two" for the first integer and "three" for the second integer:

Find the two consecutive integers such that the total of five times *three* and twice the *two* is nineteen.

The total of five times three and twice two is nineteen.

Fifteen and four is nineteen.

19 = 19 ✓

Example 4: Find three consecutive integers such that when three times the third integer is added to the sum of twice the first and the second, the result is negative eleven.

1. **Read the problem, looking for keywords.**

   Underlining the expressions before and after the AND corresponding to the leading keyword SUM OF helps avoid errors.

2. **Draw a diagram, if possible.**

   Does the problem have a total? Yes, the total is –11. Draw a board –11 long.

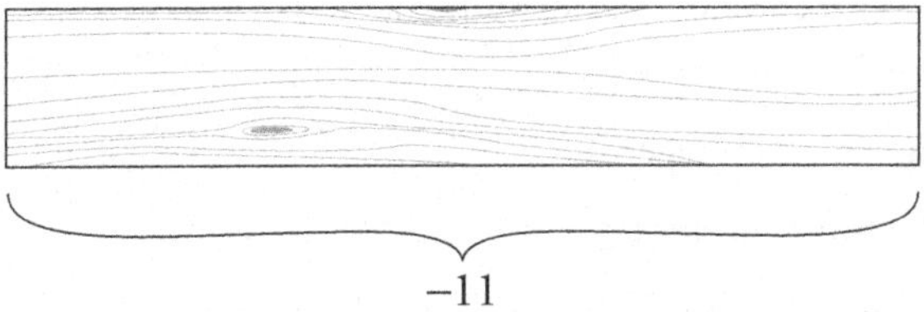

   How many things are added up to get the total of –11? Three.

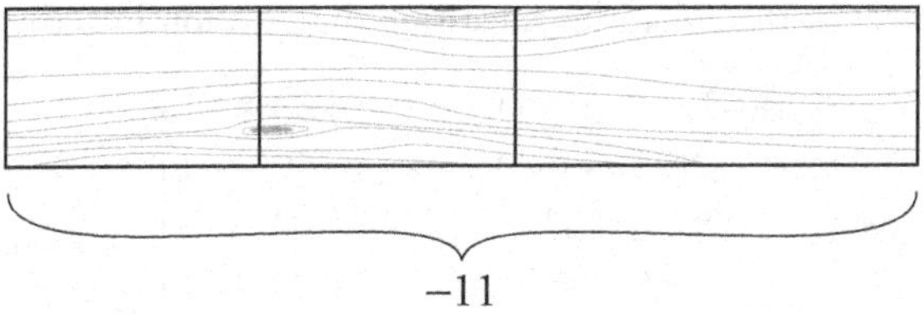

3. **Label the diagram.**

   Choose the first integer to be the unknown, x.

   The second integer is $x + 1$.

   The third integer is $x + 2$.

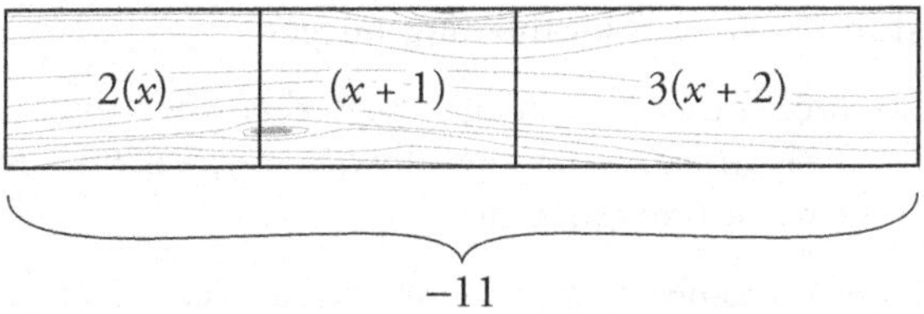

4. **Write the equation.**

$$2(x) + (x + 1) + 3(x + 2) = -11$$

5. **Solve the equation.**

$$2x + x + 1 + 3(x + 2) = -11$$

$$2x + x + 1 + 3x + 6 = -11$$

$$6x + 7 = -11$$
$$6x + 7 - 7 = -11 - 7$$
$$6x = -18$$
$$\frac{6x}{6} = \frac{-18}{6}$$
$$x = -3$$

Answer the question and include the units:

The first integer is –3.

The second integer is –2.

The third integer is –1.

Note: Integer word problems do not have units.

6. Check the solution.

a. To check the equation, substitute the solution into the equation:

$$2(-3) + (-3 + 1) + 3\,(-3 + 2\,) = -11$$
$$2(-3) + (-2) + 3\,(-3 + 2) = -11$$
$$2(-3) + (-2) + 3\,(-1) = -11$$
$$-6 + (-2) + 3\,(-1) = -11$$
$$-6 + (-2) + -3 = -11$$
$$-8 + -3 = -11$$
$$-11 = -11\ \checkmark$$

b. To check the translation, reread the question using "negative three" for the first integer, "negative two" for the second integer, and "negative one" for the third integer:

Find three consecutive integers such that when three times *negative one* is added to the sum of twice *negative three* and *negative two* the result is negative eleven.

Three times negative one is negative three.

Twice negative three is negative six.

Negative six plus negative two yields negative eight.

When you total the negative three and the negative eight, you get negative eleven.

Consecutive Even and Odd Integers

One example of three **consecutive even integers** is 6, 8, and 10. The first integer can be represented by the variable x. Notice that the second integer is not larger by one number, but by two, so it is represented by $(x + 2)$. You must add two so that you skip over the odd integer. The third even integer would then be $(x + 4)$.

Note: x, $(x + 2)$, and $(x + 4)$ also work for three consecutive even integers that are negative such as –12, –10, and –8. If x is –12, $x + 2$ is –10, and $x + 4$ is –8.

Example 5: Find two consecutive even integers such that three times the first integer plus the second even integer yields twenty-six.

1. **Read the problem, looking for keywords.**
2. **Draw a diagram, if possible.**

   Does the problem have a total? It does, so a board can be used to visualize the translation. The total is 26. Draw a board 26 long.

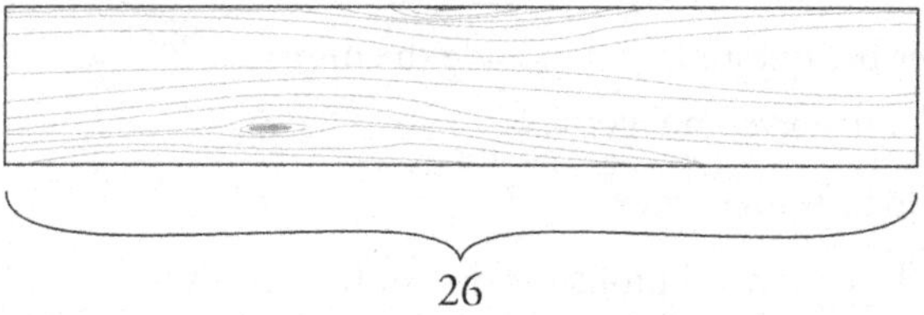

   How many things are added up to get the total of 26? Two.

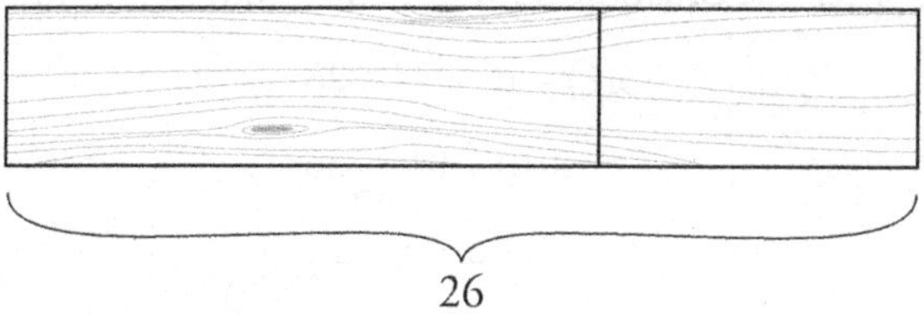

3. **Label the diagram.**

   Choose the first integer to be the unknown, x.

   The second even integer, in terms of the first, is $x + 2$.

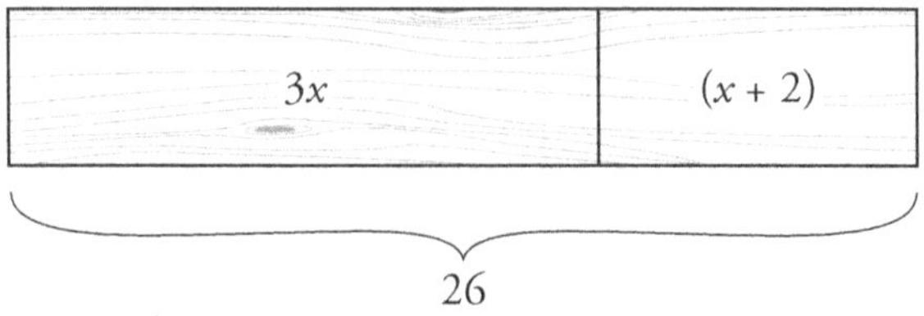

4. **Write the equation.**

$$3(x) + (x + 2) = 26$$

5. **Solve the equation.**

$$\begin{aligned} 3x + x + 2 &= 26 \\ 4x + 2 &= 26 \\ 4x + 2 - 2 &= 26 - 2 \\ 4x &= 24 \\ \frac{4x}{4} &= \frac{24}{4} \\ x &= 6 \end{aligned}$$

Answer the question and include the units:

The first even integer is 6.

The next even integer is 8.

Note: Integer word problems do not have units.

6. **Check the solution.**

a. To check the equation, substitute the solution into the equation:

$$\begin{aligned} 3(6) + (6 + 2) &= 26 \\ 3(6) + (8) &= 26 \\ 18 + 8 &= 26 \\ 26 &= 26 \checkmark \end{aligned}$$

b. To check the translation, reread the question using "six" for the first even integer and "eight" for the second even integer:

Find two consecutive even integers such that three times *six* plus *eight* yields twenty-six.

Three times six is eighteen.

Eighteen plus eight yields twenty-six.

Example 6: Find three consecutive even integers such that the sum of the integers is forty-two.

1. **Read the problem, looking for keywords.**
2. **Draw a diagram, if possible.**

   Does the problem have a total? Yes, the total is 42. Draw a board 42 long.

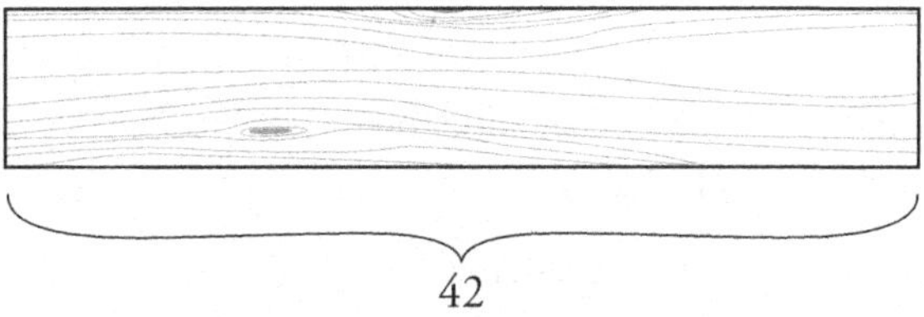

How many even integers are added up to get the total of 42? Three.

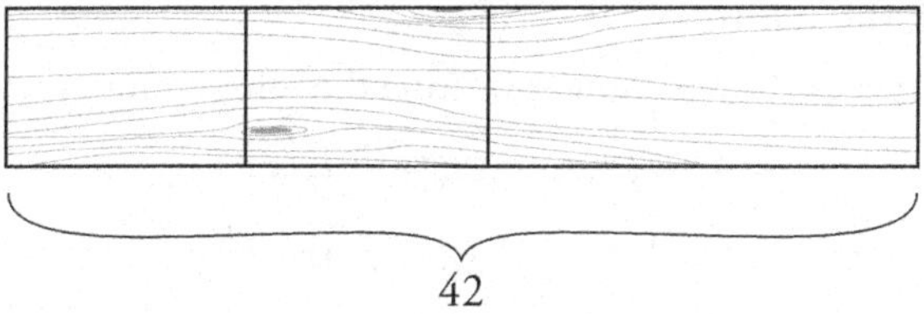

3. **Label the diagram.**

   Choose the first integer to be the unknown, x.

   The second even integer is $x + 2$.

   The third even integer is $x + 4$.

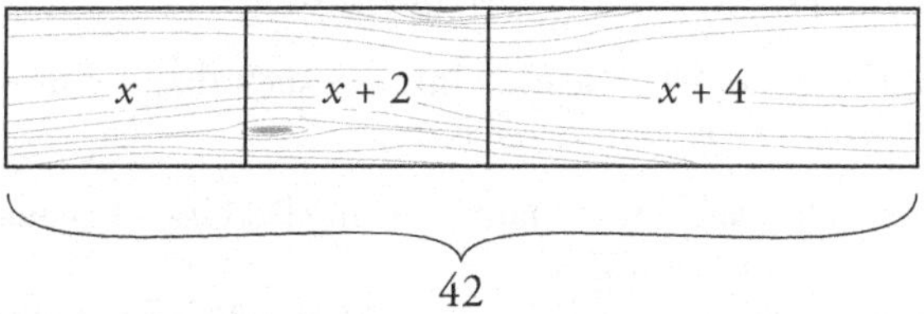

4. **Write the equation.**

$$(x) + (x + 2) + (x + 4) = 42$$

Are the parentheses necessary? No, but the parentheses do accentuate the three items being added and may help you avoid making an error.

5. Solve the equation.

$$x + x + 2 + x + 4 = 42$$
$$3x + 6 = 42$$
$$3x + 6 - 6 = 42 - 6$$
$$3x = 36$$
$$\frac{3x}{3} = \frac{36}{3}$$
$$x = 12$$

Answer the question and include the units:

The first even integer is 12.

The second even integer is 14.

The third even integer is 16.

Note: Integer word problems do not have units.

6. Check the solution.

a. To check the equation, substitute the solution into the equation:

$$12 + (12 + 2) + (12 + 4) = 42$$
$$12 + (14) + (12 + 4) = 42$$
$$12 + 14 + 16 = 42$$
$$26 + 16 = 42$$
$$42 = 42 \checkmark$$

b. To check the translation, reread the question using twelve, fourteen and sixteen for the even integers

Find three consecutive even integers such that the sum of *twelve, fourteen, and sixteen* is 42.

This is the exact same math done in checking the equation.

How do you represent **consecutive odd integers?** Consider the example of consecutive odd integers 11, 13, and 15. If x is eleven, the next odd integer is $(x + 2)$. Two must be added in order to skip over the even integer. The third consecutive odd integer is $(x + 4)$.

A common mistake made when finding consecutive odd integers is to think that the second consecutive odd integer should be $(x + 1)$. If x is 11 then $(x + 1)$ would be 12, which is not an odd integer.

Note: Whether you find consecutive odd or even integers, the unknown integers are represented by x, $(x + 2)$, and $(x + 4)$. What is the difference? The final answers should be odd or even as the original word problem states.

Example 7: Find two consecutive odd integers such that the smaller added to three times the larger gives a total of negative thirty.

1. **Read the problem, looking for keywords.**
2. **Draw a diagram, if possible.**

   Does the problem have a total? Yes, the total is –30. Draw a board –30 long.

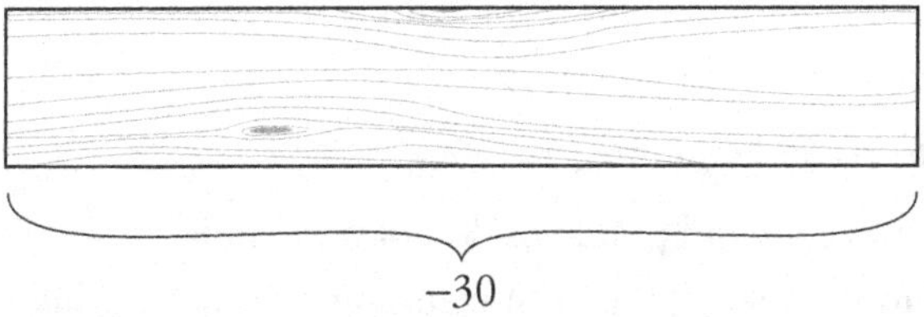

   How many things are added up to get –30? Two.

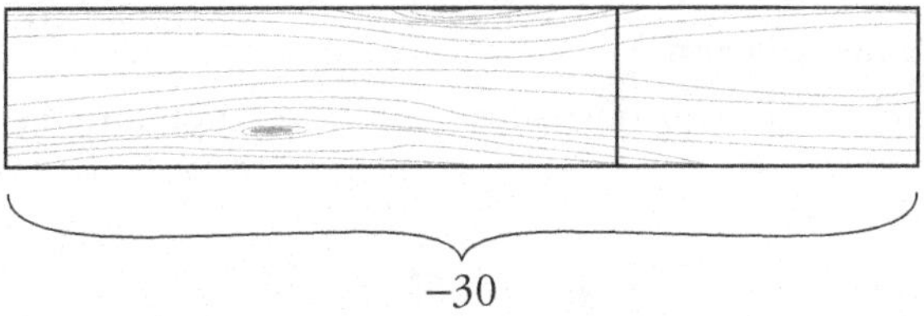

3. **Label the diagram.**

   Choose the first integer to be the unknown, x.

   The second odd integer is $x + 2$.

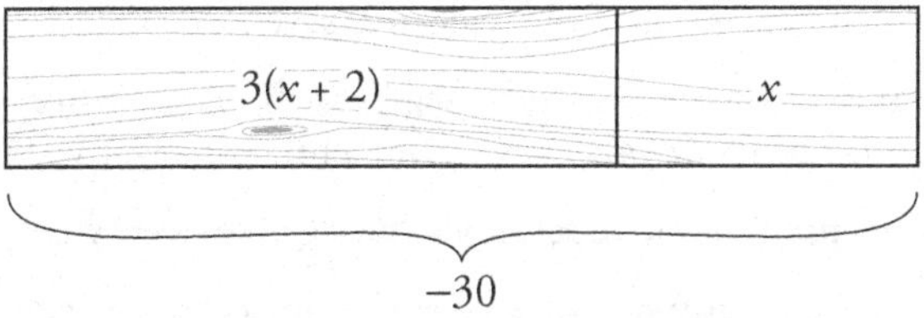

4. **Write the equation.**

$$3(x + 2) + (x) = -30$$

5. Solve the equation.

$$3(x + 2) + x = -30$$
$$3x + 6 + x = -30$$
$$4x + 6 = -30$$
$$4x + 6 - 6 = -30 - 6$$
$$4x = -36$$
$$\frac{4x}{4} = \frac{-36}{4}$$
$$x = -9$$

Answer the question and include the units:

The first odd integer is –9.

The second odd integer is –7.

Note: Integer word problems do not have units.

Remember: One of the most common mistakes on this problem is to give –11 as the second odd integer. Remember that the second odd integer can be found by adding a +2 to the –9.

6. Check the solution.

a. To check the equation, substitute the solution into the equation:

$$3(-9 + 2) + (-9) = -30$$
$$3(-7) + (-9) = -30$$
$$-21 + -9 = -30$$
$$-30 = -30 \checkmark$$

b. To check the translation, reread the question using "negative nine" for the smaller and "negative seven" for the larger:

Find two consecutive odd integers such that *negative nine* added to three times *negative seven* gives a total of negative thirty.

Three times negative seven yields negative twenty-one.

Negative twenty-one plus negative nine equals negative thirty.

Other Problems that Can Be Represented with a Board

You may be surprised just how many word problems involve a sum or total. As long as the problem has a total, even problems that do not seem easy to diagram can be represented by a board.

Example 8: In a typical small Texas town, you can find twice as many Mexican restaurants as Chinese restaurants. There is one more fried chicken restaurant than Chinese food restaurants. In a town with a total of thirteen Mexican, Chinese, and fried chicken restaurants, how many of each type are in the town?

1. **Read the problem, looking for keywords.**
2. **Draw a diagram, if possible.**

   Does the problem have a total? Yes, the total is 13 restaurants. Draw a board 13 restaurants long.

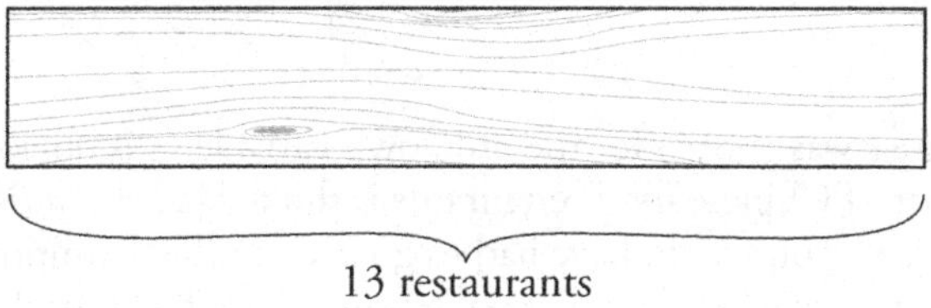

   How many things are added up to get the 13 restaurants? Three.

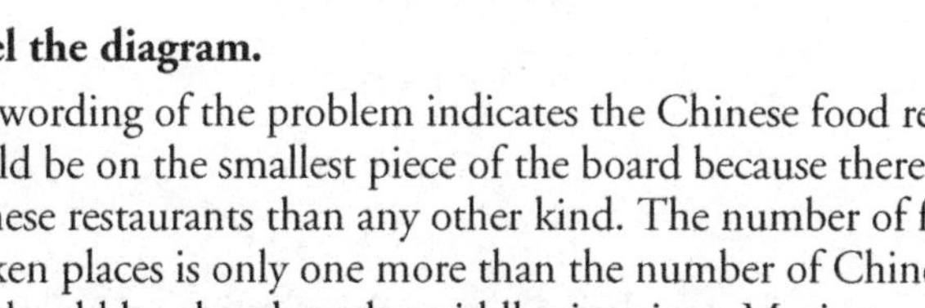

3. **Label the diagram.**

   The wording of the problem indicates the Chinese food restaurants should be on the smallest piece of the board because there are fewer Chinese restaurants than any other kind. The number of fried chicken places is only one more than the number of Chinese places and should be placed on the middle-size piece. Mexican restaurants should be placed on the largest piece of board.

   Choose the unknown to be c for Chinese because all the other restaurants are compared to the number of Chinese food restaurants.

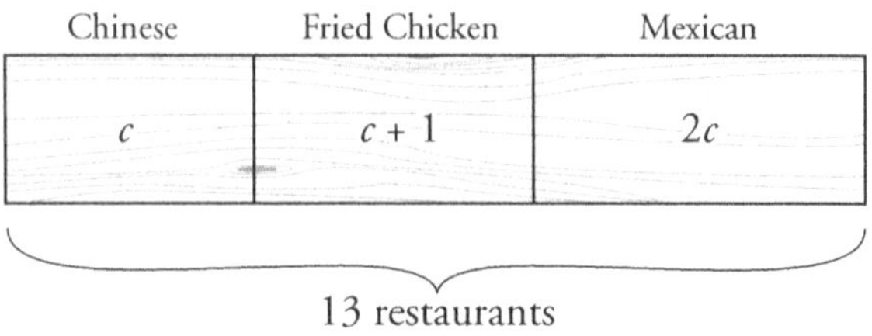

4. **Write the equation.**

$$c + (c + 1) + (2c) = 13$$

5. **Solve the equation.**

$$c + c + 1 + 2c = 13$$
$$4c + 1 = 13$$
$$4c + 1 - 1 = 13 - 1$$
$$4c = 12$$
$$\frac{4c}{4} = \frac{12}{4}$$
$$c = 3$$

Because c was chosen for the unknown variable, it is obvious the number of Chinese food restaurants is three. Had the variable x been used, you would have had to go back to the definition of the variable to be sure which type of restaurant we had found.

The number of Chinese food restaurants is 3; there are 4 fried chicken places, and 6 Mexican food restaurants.

6. **Check the solution.**

a. To check the equation, substitute the solution into the equation:

$$3 + (3 + 1) + 2(3) = 13$$
$$3 + 4 + 2(3) = 13$$
$$3 + 4 + 6 = 13$$
$$7 + 6 = 13$$
$$13 = 13 \checkmark$$

b. To check the translation, reread the question using "three" for the number of Chinese food restaurants:

In a typical small Texas town, you can find twice as many Mexican food restaurants as three. There is one more fried chicken restaurant than three. In a town with a total of thirteen Mexican, Chinese, and fried chicken restaurants, how many of each type are in the town?

The number of Mexican food restaurants is twice three, which is six.

One more than three is four fried chicken restaurants.

If you add three, four, and six, you get a total of thirteen restaurants.

Example 9: Jo Anna has two antennae on her roof, broadband Internet and satellite television. The satellite television has a monthly flat fee that is twice the fee for the broadband, plus a $15 fee for extra channels. The total monthly cost of the two bills is $74.85. How much is the monthly rate for broadband?

1. **Read the problem, looking for keywords.**
2. **Draw a diagram, if possible.**

   Does the problem have a total? Yes, the total is $74.85. Draw a board $74.85 long.

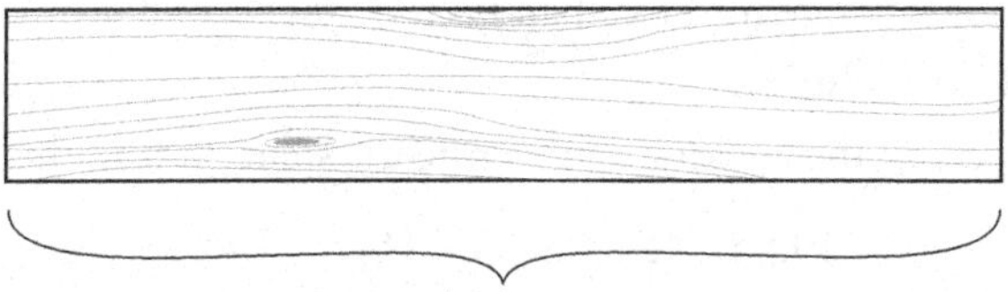

$74.85

How many things are added up to get $74.85? Two.

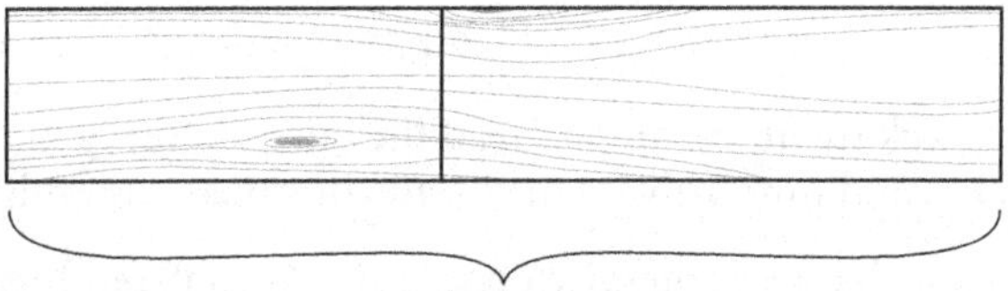

$74.85

3. **Label the diagram.**

   Choose the unknown to be *b* (for broadband) because the satellite charges are compared to the charges for broadband.

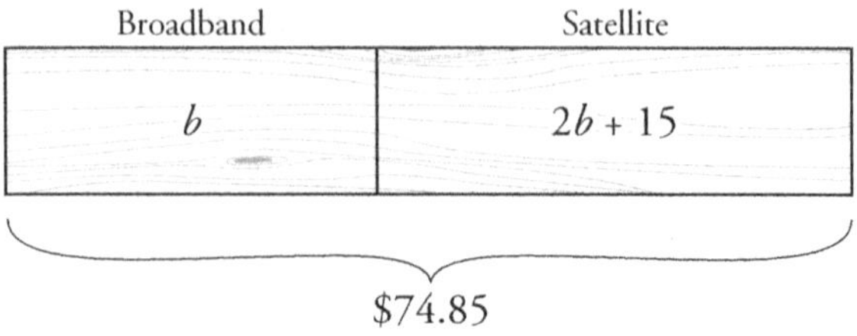

4. Write the equation.

$$b + (2b + 15) = 74.85$$

5. Solve the equation.

$$3b + 15 = 74.85$$
$$3b + 15 - 15 = 74.85 - 15$$
$$3b = 59.85$$
$$\frac{3b}{3} = \frac{59.85}{3}$$
$$b = 19.95$$

Because b was chosen for the unknown variable, the charge for broadband is $19.95.

The monthly rate for broadband is $19.95.

6. Check the solution.

a. To check the equation, substitute the solution into the equation:

$$19.95 + (2 \times 19.95 + 15) = 74.85$$
$$19.95 + (39.90 + 15) = 74.85$$
$$19.95 + (54.90) = 74.85$$
$$74.85 = 74.85 \checkmark$$

b. To check the translation, reread the question using nineteen dollars and ninety-five cents for the broadband monthly rate:

Jo Anna has two antennae on her roof, broadband Internet and satellite television. The satellite television has a monthly flat fee that is twice $19.95, plus a $15 fee for extra channels. The total monthly cost of the two bills is $74.85. How much is the monthly rate for broadband?

The monthly fee for satellite is twice $19.95 plus $15.

Twice $19.95 is $39.90.

When you add \$15 and \$39.90, you get a total of \$54.90 for the satellite.

The satellite bill plus the broadband bill is \$74.85.

One common type of word problem not obviously easy to diagram is called a work problem. A work problem can be represented by a board if the job is considered a total of one, as shown in Examples 10 and 11.

Example 10: Bita and Martha want to paint their living room. If Bita painted it by herself, it would take her six hours. If Martha painted the room herself, it would take eight hours. How long will it take them to paint the room if they work together?

1. **Read the problem, looking for keywords.**
2. **Draw a diagram, if possible.**

   This problem does have a total, but it is a bit more difficult to ascertain. The total room will be painted (the whole job) so the total for the board is 1.
3. **Label the diagram.**

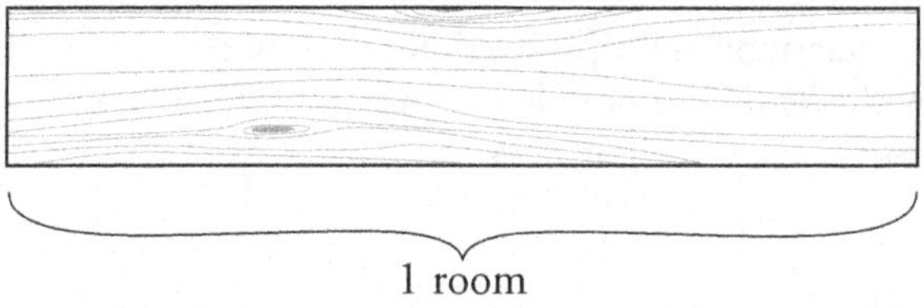

How many things are added up to get the whole job? Two.

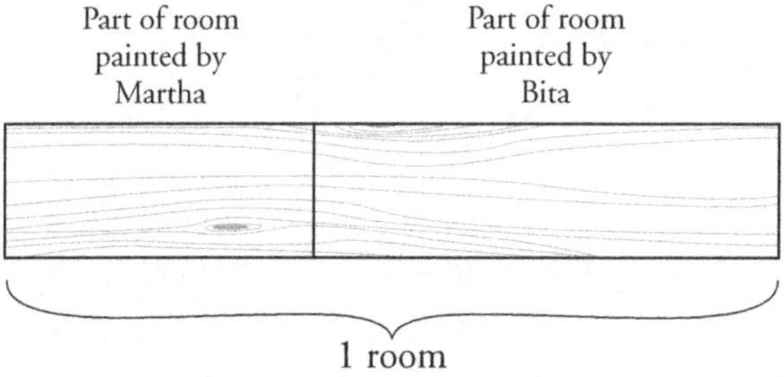

Note: Because Martha paints slower than Bita, she will paint less of the room.

The unknown in this problem is the amount of time it will take them to paint the room. To find how much of the room Martha will paint, multiply the number of hours painting by her speed.

Martha can paint the room in eight hours. How much of the room will she paint in one hour? $\frac{1}{8}$. Her speed is therefore $\frac{1}{8}$. What is Bita's rate? Bita's speed is one room per six hours, or $\frac{1}{6}$.

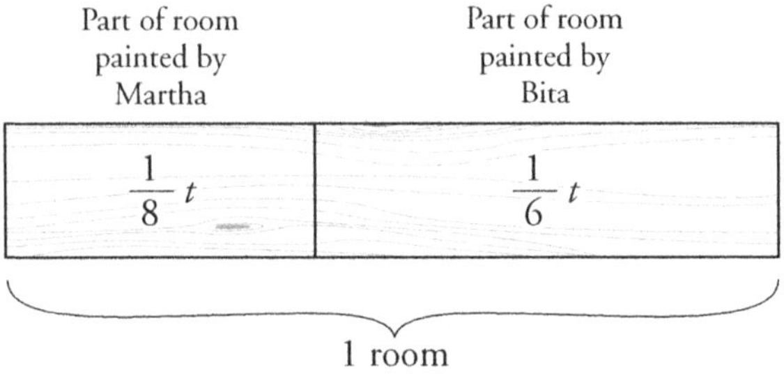

Remember: The portion of room painted by Martha is her rate times the number of hours she paints.

4. **Write the equation.**

$$\frac{1}{8}t+\frac{1}{6}t=1$$

5. **Solve the equation.**

To clear fractions, multiply every term on both sides of the equation by the least common denominator, LCD. The **least common multiple** of six and eight is 24.

$$24\left(\frac{1}{8}t+\frac{1}{6}t\right)=24(1)$$

$$24\left(\frac{1}{8}t+\frac{1}{6}t\right)=24$$

$$24\left(\frac{1}{8}\right)t+\frac{24}{1}\left(\frac{1}{6}\right)t=24$$

$$\frac{\overset{3}{\cancel{24}}}{1}\left(\frac{1}{\underset{1}{\cancel{8}}}\right)t+\frac{\overset{4}{\cancel{24}}}{1}\left(\frac{1}{\underset{1}{\cancel{6}}}\right)t=24$$

$$3t+4t=24$$

$$7t=24$$

$$\frac{7t}{7}=\frac{24}{7}$$

$$t=3\frac{3}{7}$$

Answer the question and include the units with your answer:
They can paint the room in $3\frac{3}{7}$ hours if they work together.

6. **Check the solution.**

   Many people skip this step if the solution involves fractions. Two methods of checking are given below to show the ease of checking, even with fractions.

   a. Estimating: Does the answer seem reasonable? If Bita were cloned, she and her clone could paint the room in three hours. If Martha and her clone painted the room it would take four hours. $3\frac{3}{7}$ is between three and four hours, so it is a reasonable answer.

   b. Checking the solution in the equation:

$$\frac{1}{8}\left(\frac{24}{7}\right)+\frac{1}{6}\left(\frac{24}{7}\right)=1$$

$$\frac{1}{\cancel{8}_1}\left(\frac{\cancel{24}^3}{7}\right)+\frac{1}{\cancel{6}_1}\left(\frac{\cancel{24}^4}{7}\right)=1$$

$$\frac{3}{7}+\frac{4}{7}=1$$

$$\frac{7}{7}=1$$

$$1=1\ \checkmark$$

Example 11: The Banner Press has purchased a new printing press. The new press can print all the Sunday papers in three hours. The old press still works and has always taken five hours to print all the Sunday papers. To save printing time, the Banner Press has decided to use both presses for this Sunday paper. How long will it take for the Sunday paper to be printed?

1. **Read the problem, looking for keywords.**
2. **Draw a diagram, if possible.**

   The total of all the Sunday papers to be printed is the whole job, so the total for the board is 1.

3. Label the diagram.

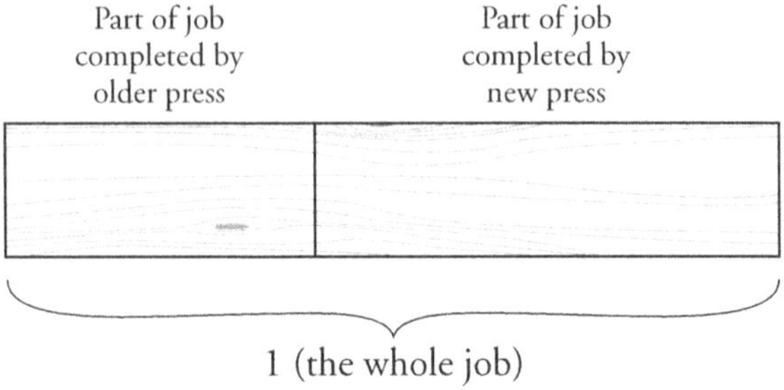

The unknown in this problem is the amount of time it will take both presses working at the same time to print all of the Sunday papers. To find how much of the job will be printed by the older press, multiply the number of hours printing by the speed of the older press.

The speed of the older press is measured in part of job per hour. The older press is able to do $\frac{1}{5}$ of the job in one hour. The speed of the new press is $\frac{1}{3}$.

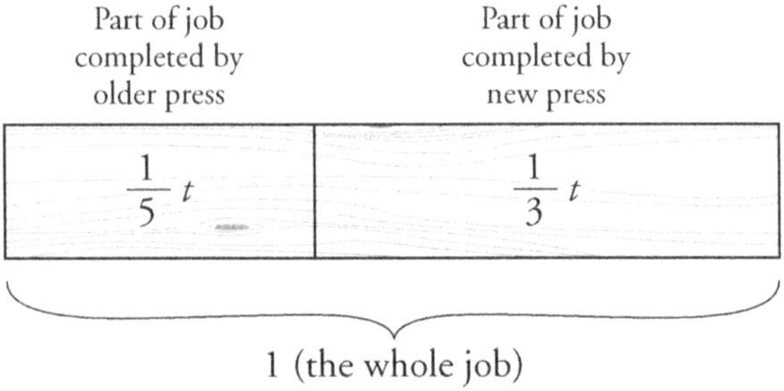

Remember: The portion of the job printed by the older press is its hourly rate times the number of hours printing.

4. Write the equation.

$$\frac{1}{5}t+\frac{1}{3}t=1$$

5. Solve the equation.

To clear fractions, multiply every term on both sides of the equation by the least common denominator, LCD. The least common multiple of five and three is fifteen.

$$15\left(\frac{1}{5}t+\frac{1}{3}t\right)=15(1)$$

$$15\left(\frac{1}{5}t + \frac{1}{3}t\right) = 15$$

$$\frac{15}{1}\left(\frac{1}{5}\right)t + \frac{15}{1}\left(\frac{1}{3}\right)t = 15$$

$$\frac{\overset{3}{\cancel{15}}}{1}\left(\frac{1}{\underset{1}{\cancel{5}}}\right)t + \frac{\overset{5}{\cancel{15}}}{1}\left(\frac{1}{\underset{1}{\cancel{3}}}\right)t = 15$$

$$3t + 5t = 15$$

$$8t = 15$$

$$\frac{8t}{8} = \frac{15}{8}$$

$$t = 1\frac{7}{8}$$

Answer the question and include the units with your answer: The Banner Press can print the Sunday paper in $1\frac{7}{8}$ hours if it uses both presses.

6. **Check the solution.**

   a. Estimation: Does the answer seem reasonable? If two identical new presses were used, the Sunday paper could be printed in one and a half hours. If two identical older presses were used, the Sunday paper could be printed in two and a half hours. $1\frac{7}{8}$ is between both of those times.

   b. Checking the solution in the equation:

$$\frac{1}{5}\left(\frac{15}{8}\right) = \frac{1}{3}\left(\frac{15}{8}\right) = 1$$

$$\frac{1}{\underset{1}{\cancel{5}}}\left(\frac{\overset{3}{\cancel{15}}}{8}\right) + \frac{1}{\underset{1}{\cancel{3}}}\left(\frac{\overset{5}{\cancel{15}}}{8}\right) = 1$$

$$\frac{3}{8} + \frac{5}{8} = 1$$

$$\frac{8}{8} = 1$$

$$1 = 1 \checkmark$$

Chapter Checkout

Q&A

Solve the following problems. Be sure to answer the question and include the units.

1. Kevin wants to build an open box to use for storage from a scrap 1 x 10 board that is 6'2" long. He has drawn a rough sketch of the box (see the following figure). The two ends must be 8.5 inches long. The sides and bottom are all the same length. How long are the sides and bottom?

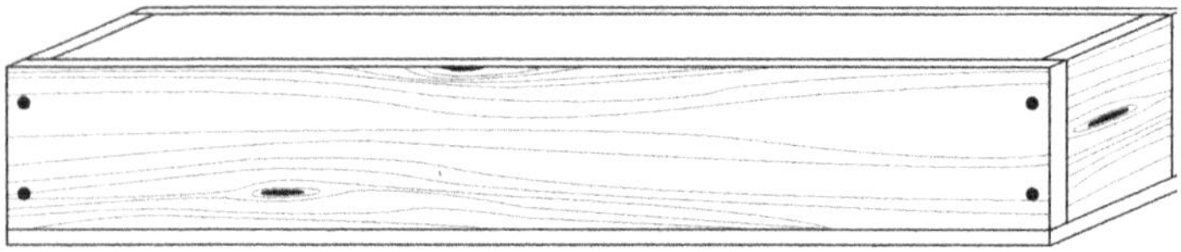

2. A woodworking project requires three pieces of board. Kameron has an eight-foot board. The longest piece must be twice the length of the smallest piece. The middle-size piece must be four inches shorter than the longest piece. Find the length of all three pieces.
3. Find three consecutive integers such that the sum of the integers is 36.
4. Find two consecutive integers such that seven times the first integer plus the second is 81.
5. Find two consecutive odd integers such that the sum of the first and twice the second is 55.
6. Find three consecutive even integers such that the sum of the first, third, and twice the second is –16.
7. Find three consecutive odd integers such that when five times the third minus the second is added to the first, the result is 23.
8. The tiniest dog in the world is a saucy Chihuahua. The average dachshund is one pound more than twice the weight of an average Chihuahua. An average-size collie weighs fifteen times more than an average Chihuahua. If the total weight of the three average dogs is 91 pounds, what is the weight of each dog?
9. For every one hundred people in the world, the world also has a total of 207 pigs, camels, and chickens. There are five times as many pigs as camels and sixty-three times as many chickens as camels. How many of each type of animal are there for every one hundred people in the world?

10. Howard has a collection of ceramic frogs. Angie collects teapots. Howard has twice as many frogs as Angie has teapots. Together they have a total of 177 collectables. How many teapots does Angie have in her collection?

11. Dotti and Kenneth both stuff envelopes for the electric company. Dotti can stuff all the electric company bills into envelopes in 4 hours. Kenneth can stuff the bills in 5 hours. How long would it take them if they worked together?

12. A large bakery in New York has two machines it uses to mix the dough for doughnuts. The faster machine can make enough dough for a Monday morning in 2 hours. The slower machine takes 3 hours to make the same amount of dough. How long will it take to make enough doughnut dough for a Monday morning if both machines are used?

Answers: 1. The sides and the bottom are all 19 inches long. **2.** The shortest piece is 20 inches long; the middle-size piece is 36 inches long; the longest piece is 40 inches long. **3.** The three consecutive integers are 11, 12, and 13. **4.** The two consecutive integers are 10 and 11. **5.** The two consecutive odd integers are 17 and 19. **6.** The three consecutive even integers are –6, –4, and –2. **7.** The three consecutive odd integers are 1, 3, and 5. **8.** The weight of the Chihuahua is 5 pounds; the weight of the dachshund is 11 pounds; the weight of the collie is 75 pounds. **9.** There are 3 camels, 15 pigs, and 189 chickens for every 100 people in the world. **10.** Angie has 59 teapots in her collection. **11.** It would take $2\frac{2}{9}$ hours for Dotti and Kenneth to stuff the bills together. **12.** It will take $1\frac{1}{5}$ hours to make enough doughnut dough.

Chapter 8

SOLVING WORD PROBLEMS USING TRIED-AND-TRUE METHODS

Chapter Check-In

- ❑ Polya's four-step process and its evolution
- ❑ Consecutive integers
- ❑ Consecutive even and odd integers
- ❑ Other problems that cannot be represented with a board

Not all word problems can be solved using the board method, as discussed in Chapter 7. While word problems are usually easiest to visualize and solve when they're diagramed, sometimes, creating a diagram is not an option. Other tools, which are discussed in this chapter, are available to help you translate the correct equation from a word problem. The **tried-and-true methods** presented in this chapter have been developed over many years. These methods have proven successful when solving word problems, especially those which cannot be diagramed.

Polya's Four-Step Process

George Polya was a professor at Stanford University who studied problem-solving techniques. In 1945, he published his four-step process of problem solving:

1. **Understand the problem.**
2. **Devise a plan.**
3. **Carry out the plan.**
4. **Look back over the results.**

These steps are the foundation for the six steps for solving word problems given in Chapter 5. Polya's four steps are rewritten as follows:

1. **Read the problem. (Understand the problem.)**
2. **Write an equation. (Devise a plan.)**
3. **Solve the equation. (Carry out the plan.)**
4. **Check. (Look back over the results.)**

Expanding on Polya's four steps

Over the years, additional successful techniques have been created, but all presentations of problem-solving strategies are based on Polya's four original steps. This section shows each of Polya's steps with additional, more specific techniques.

1. **Understand the problem.**
   - **a.** Read the problem.
   - **b.** Recognize keywords that indicate operations and parentheses.
   - **c.** Recognize extraneous information.
   - **d.** Identify the unknown variable(s). Assign the variable a letter of the alphabet that makes sense and write a brief description of the variable.
2. **Devise a plan.**
   - **a.** Draw and label a picture.
   - **b.** Estimate the value(s) of the variable(s).
   - **c.** Translate an English sentence into an algebraic equation.
   - **d.** Write an equation.
3. **Carry out the plan.**
   - **a.** Solve the equation(s).
4. **Look back over the results.**
   - **a.** Reread the word problem to ensure that you have answered the question asked.
   - **b.** Check the solution of the equation.
   - **c.** Check the translation of the word problem.
   - **d.** Be sure your solution is reasonable by comparing it to your estimate.

The five steps of the tried-and-true methods

Because the list in the preceding section is too detailed to memorize, focus on the following list of steps:

1. **Read the problem.**

   Make helpful markings for turnaround words and leading keywords as you read the problem. (See Chapters 1 and 2.)

2. **Identify the variable(s).**

   Look at the question to determine what value(s) are to be found. Write down the variable that represents the unknown value and include a brief description. If possible, choose a letter of the alphabet that reflects the unknown quantity. Estimate the solution.

3. **Translate the equation(s).**

   Reread the problem one sentence at a time. Translate each sentence into an equation, if possible. You may have to gather information from other parts of the word problem in order to write the equation(s).

4. **Solve the equation(s).**

   Simple linear equations can be solved by methods covered in Chapter 4. When a problem has more than one equation, use the substitution method or elimination method of solving systems of equations, discussed in Chapter 10.

5. **Check the solution.**

   Reread the question to be sure it was answered. Include units on your answer. Ask yourself whether the answer is reasonable. Check your translation.

Consecutive Integers

Chapter 7 shows you how to set up and solve consecutive integer problems. Examples 1 and 2 in this chapter are presented as Examples 3 and 4 in Chapter 7, where the board method is used to solve them. Here, however, tried-and-true methods are used to solve the word problems. Either method can be used to successfully solve these problems.

Example 1: Find two consecutive integers such that the total of five times the second integer and twice the first integer is nineteen.

1. **Read the problem, making helpful markings for the leading keyword.**

Find the two consecutive integers such that the

(total of) five times the second integer and

twice the first integer is nineteen.

Note: Underline the expressions before and after the AND that corresponds to the leading keyword TOTAL OF. Remember that IS not only indicates inequality, but also acts as a separator. (See Chapters 1&2 for a review of keywords.)

2. **Identify the variables.**

x, the first integer

$x + 1$, the second integer

3. **Translate the equation.**

$$5(x + 1) + 2(x) = 19$$

4. **Solve the equation.**

$$5(x + 1) + 2x = 19$$
$$5x + 5 + 2x = 19$$
$$7x + 5 = 19$$
$$7x + 5 - 5 = 19 - 5$$
$$7x = 14$$
$$\frac{7x}{7} = \frac{14}{7}$$
$$x = 2$$

Answer the question and include the units:

The first integer is 2.

The second integer is 3.

Note: Integer word problems do not have units.

5. **Check the solution.**

Reread the question using "two" for the first integer and "three" for the second integer:

Find the two consecutive integers such that the total of five times *three* and twice the *two* is nineteen.

Five times three is fifteen.

Twice two is four.

When you find the total of fifteen and four you get nineteen. ✓

Example 2: Find three consecutive integers such that when three times the third integer is added to the sum of twice the first and the second, the result is negative eleven.

1. **Read the problem, making helpful markings.**

   Find three consecutive integers such that when

   (three times the third integer) is added to

   the sum of twice the first and the second,

   the result is negative eleven.

2. **Identify the variables.**

   x, the first integer
   $x + 1$, the second integer
   $x + 2$, the third integer

3. **Translate the equation.**

   $$2(x) + (x + 1) + 3(x + 2) = -11$$

4. **Solve the equation.**

   $$2x + x + 1 + 3(x + 2) = -11$$
   $$2x + x + 1 + 3x + 6 = -11$$
   $$6x + 7 = -11$$
   $$6x + 7 - 7 = -11 - 7$$
   $$6x = -18$$
   $$\frac{6x}{6} = \frac{-18}{6}$$
   $$x = -3$$

   Answer the question and include the units:

   The first integer, x, is –3.

   The second integer, $x + 1$, is –2.

   The third integer, $x + 2$, is –1.

Note: Integer word problems do not have units.

5. **Check the solution.**

   a. To check the equation, substitute the solution into the equation:

$$2(-3) + (-3 + 1) + 3\,(-3 + 2) = -11$$
$$2(-3) + (-2) + 3\,(-3 + 2) = -11$$
$$2(-3) + (-2) + 3\,(-1) = -11$$
$$-6 + (-2) + 3\,(-1) = -11$$
$$-6 + (-2) + -3 = -11$$
$$-8 + -3 = -11$$
$$-11 = -11 \checkmark$$

   b. To check the translation, reread the question using "negative three" for the first integer, "negative two" for the second integer, and "negative one" for the third integer:

   Find three consecutive integers such that when three times *negative one* is added to the sum of twice *negative three* and *negative two,* the result is negative eleven.

   Three times negative one is negative three.

   The sum of twice negative three and negative two is negative six plus negative two, which yields negative eight.

   When you total the negative eight and the negative three, you get negative eleven. ✓

Example 3 does not include a total and, thus, cannot be solved using the board method discussed in Chapter 7. Use the tried-and-true methods, as shown in the explanation following Example 3.

Example 3: Find three consecutive integers such that the sum of the first integer and the second integer is equal to the third integer.

1. **Read the problem, making helpful markings.**

   Find three consecutive integers such that the

   sum of the first and the second integer is

   equal to the third integer.

Note: Underline the expressions before and after the AND that corresponds to the leading keyword SUM OF.

2. **Identify the variables.**

$$x\text{, the first integer}$$
$$x + 1\text{, the second integer}$$
$$x + 2\text{, the third integer}$$

3. **Translate the equation.**

$$x + (x + 1) = (x + 2)$$

4. **Solve the equation.**

$$2x + 1 = x + 2$$
$$2x + 1 - x = x + 2 - x$$
$$x + 1 = 2$$
$$x + 1 - 1 = 2 - 1$$
$$x = 1$$

Answer the question and include the units:

The first integer is 1.

The second integer is 2.

The third integer is 3.

Note: Integer word problems do not have units.

5. **Check the solution.**

Reread the question using "one" for the first integer, "two" for the second integer, and "three" for the third integer:

Find three consecutive integers such that the sum of *one* and *two* is equal to *three.*

$$1 + 2 = 3$$
$$3 = 3 \checkmark$$

Consecutive Even and Odd Integers

In Example 4, you practice using the tried-and-true methods for even and odd integer problems, which are discussed more fully in Chapter 7.

Example 4: Find two consecutive even integers such that four times the first integer is equal to the second even integer plus ten.

1. **Read the problem.**
2. **Identify the variables.**

x, the first even integer
$x + 2$, the second even integer

3. **Write the equation.**

$$4(x) = (x + 2) + 10$$

4. **Solve the equation.**

$$4x = x + 12$$
$$4x - x = x + 12 - x$$
$$3x = 12$$
$$\frac{3x}{3} = \frac{12}{3}$$
$$x = 4$$

Answer the question and include the units:

The first even integer is 4.

The next even integer is 6.

Note: Integer word problems do not have units.

5. **Check the solution.**

a. To check the equation, substitute the solution into the equation:

$$4(4) = (4 + 2) + 10$$
$$4(4) = (6) + 10$$
$$16 = 16 \checkmark$$

b. To check the translation, reread the question using "four" for the first even integer and "six" for the second even integer:

Find two consecutive even integers such that four times *four* is equal to *six* plus ten.

Four times four is sixteen.

Six plus ten equals sixteen. ✓

Example 5: Find three consecutive odd integers such that the sum of the first odd integer and the second odd integer is equal to four times the third.

1. **Read the problem, making helpful markings.**

Find three consecutive odd integers such that the sum of the first odd integer and the second odd integer is equal to four times the third.

2. **Identify the variables.**

x, the first odd integer
$x + 2$, the second odd integer
$x + 4$, the third odd integer

3. **Write the equation.**

$$x + (x + 2) = 4(x + 4)$$

4. **Solve the equation.**

$$x + x + 2 = 4(x + 4)$$
$$2x + 2 = 4x + 16$$
$$2x + 2 - 4x = 4x + 16 - 4x$$
$$-2x + 2 = 16$$
$$-2x + 2 - 2 = 16 - 2$$
$$-2x = 14$$
$$\frac{-2x}{-2} = \frac{14}{-2}$$
$$x = -7$$

Answer the question and include the units:

The first odd integer, x, is –7.

The second odd integer, $x + 2$, is –5.

The third odd integer, $x + 4$, is –3.

Note: Integer word problems do not have units.

5. **Check the solution.**

a. To check the equation, substitute the solutions into the equation:

$$-7 + (-7 + 2) = 4(-7 + 4)$$
$$-7 + (-5) = 4(-3)$$
$$-12 = -12 \checkmark$$

b. To check the translation, reread the question using "negative seven" for the first odd integer, "negative five" for the second odd integer, and "negative three" for the third odd integer:

Find three consecutive odd integers such that the sum of *negative seven* and *negative five* is equal to four times *negative three.*

The sum of negative seven and negative five is negative twelve.

Four times negative three is negative twelve. ✓

Other Problems that Cannot Be Represented with a Board

Follow the five steps of the tried-and-true methods to solve Example 6.

Example 6: In 1821, the first tunnel of any kind was dug in the United States — the 450-foot-long tunnel was dug in Orwigsburg, Pennsylvania. Now, many tunnels crisscross the United States. The longest tunnel is the Brooklyn-Battery in New York: It is 117 feet longer than twenty times the first tunnel in the United States. Find the length of the Brooklyn-Battery tunnel.

1. **Read the problem.**
2. **Identify the variables.**

   b, the length of the Brooklyn-Battery tunnel

3. **Translate an equation.**

   The sentence that will translate into the required equation is "It is 117 feet longer than twenty times the first tunnel in the United States." "It" refers to the longest tunnel, which is the unknown, b. The length of "the first tunnel in the United States" is 450 feet, as given in the problem.

   $$b = 20 \times 450 + 117$$

4. **Solve the equation.**

   $$b = 9{,}000 + 117$$
   $$b = 9{,}117$$

   Answer the question and include the units: The length of the Brooklyn-Battery tunnel is 9,117 feet long.

5. **Check the solution.**

   Reread the question using "9,117" for the length of the Brooklyn-Battery tunnel and "450" for the length of the first U.S. tunnel:

   9,117 is 117 feet longer than twenty times *450.*

   Twenty times 450 is 9,000.

   Add 117 to 9,000, and the result is 9,117. ✓

Example 7: According to the 2000 census, the United States has many Native American tribes. The largest tribe is the Cherokee. The Cherokee tribe has 1,318 more members than 25 times the number of members of the Cheyenne tribe. Of the people who took the 2000 census, 11,191 were members of the Cheyenne tribe. How many members of the Cherokee tribe took the census?

1. **Read the problem, making helpful markings.**

   The Cherokee tribe has 1,318 more members

   [than] 25 times the number of members of

   the Cheyenne tribe.

2. **Identify the variables.**

   x, the number of members of the Cherokee tribe

   Note: Using the variable *c* is not advisable in this problem because both Cheyenne and Cherokee begin with the letter "c."

3. **Translate an equation.**

   $$x = 25 \times 11{,}191 + 1{,}318$$

4. **Solve the equation.**

   $$x = 279{,}775 + 1{,}318$$
   $$x = 281{,}093$$

   Answer the question and include the units: The Cherokee tribe has 281,093 members who took the 2000 Census.

5. **Check the solution.**

   Reread the question using "281,093" for the number of members of the Cherokee tribe and "11,191" for the number of members of the Cheyenne tribe:

   281,093 is 1,318 more members than 25 times *11,191.*

$$281{,}093 = 25 \times 11{,}191 + 1{,}318$$
$$281{,}093 = 279{,}775 + 1{,}318$$
$$281{,}093 = 281{,}093 \checkmark$$

Example 8: In 1941, Joe DiMaggio set a record in baseball by having at least one hit in 56 consecutive games. His second year as a Yankee, he made 17 more home runs than in the 1936 season as a rookie. He made 46 home runs in his second year. Find the number of home runs he made as a rookie.

1. **Read the problem, making helpful markings.**

   His second year as a Yankee, he made
   seventeen more home runs [than] in the
   1936 season as a rookie.

   Remember: The information about the 56 consecutive games is extraneous.

2. **Identify the variables.**

   r, the number of home runs made as a rookie

3. **Translate an equation.**

   $$46 = r + 17$$

   Use estimation to confirm this equation. Joe DiMaggio improved from his first year to his second year; therefore, he added home runs to his rookie year, *r*.

4. **Solve the equation.**

   $$46 - 17 = r + 17 - 17$$
   $$29 = r$$

   Answer the question and include the units: Joe DiMaggio made 29 home runs his first year as a Yankee.

5. **Check the solution.**

   Reread the question using 29 for the number of home runs his first year and 46 his second year.

   Forty-six is seventeen more home runs than *29.*

   $$46 = 29 + 17$$
   $$46 = 46 \checkmark$$

Example 9: A favorite honeymoon spot is the Horseshoe Falls of Niagara. While Niagara Falls are beautiful, the highest waterfall is Angel Falls in Venezuela. The water at Angel Falls drops 3,212 feet. The difference between the height of Angel Falls and the height of the Horseshoe Falls is seventeen times the height of the Horseshoe Falls, plus 98 feet. Find the height of the Horseshoe Falls.

1. **Read the problem, making helpful markings.**

   The difference between the height of Angel Falls and the height of the Horseshoe Falls is seventeen times the height of the Horseshoe Falls, plus 98 feet.

2. **Identify the variable.**

   h, the height of Horseshoe Falls

3. **Translate an equation.**

$$3{,}212 - h = 17h + 98$$

4. **Solve the equation.**

$$3{,}212 - h - 17h = 17h + 98 - 17h$$
$$3{,}212 - 18h = 98$$
$$3{,}212 - 18h - 3{,}212 = 98 - 3{,}212$$
$$-18h = -3{,}114$$
$$\frac{-18h}{-18} = \frac{-3,114}{-18}$$
$$h = 173$$

   Answer the question and include the units: The height of Horseshoe Falls is 173 feet.

5. **Check the solution.**

   a. To check the equation, substitute the solutions into the equation:

$$3{,}212 - 173 = 17(173) + 98$$
$$3{,}212 - 173 = 2{,}941 + 98$$
$$3{,}039 = 3{,}039 \checkmark$$

b. To check the translation, reread the question using "173" for the height of Horseshoe Falls and "3,212" for the height of Angel Falls:

The difference between *3,212* and *173* is seventeen times *173* plus 98 feet.

The difference between 3,212 and 173 is 3,039.
Seventeen times 173 is 2,941.
2,941 plus 98 is 3,039. ✓

Follow the five steps of the tried-and-true methods to solve the following work problems. As discussed in Chapter 7, work problems total to one whole job.

Example 10: Martin and Nabil have a lawn-maintenance business. When Nabil maintains Mrs. Bilski's lawn, it takes him two hours to complete the job. Martin takes three hours to do the same lawn. How long will it take them to maintain Mrs. Bilski's lawn if they work together?

1. **Read the problem.**
2. **Identify the variables.**

   t, the time it takes to maintain the lawn working together

3. **Write the equation.**

   Find the portion of the lawn Martin will maintain in one hour. His speed is therefore $\frac{1}{3}$. What is Nabil's rate? Nabil's speed is one lawn per two hours, or $\frac{1}{2}$.
   The portion Martin maintains plus the portion Nabil maintains add up to the whole job, 1.

$$\frac{1}{3}t+\frac{1}{2}t=1$$

4. **Solve the equation.**

   To clear fractions, multiply every term on both sides of the equation by the **least common denominator** (LCD). The **least common multiple** of 3 and 2 is 6.

$$6\left(\frac{1}{3}t+\frac{1}{2}t\right)=6$$

$$6(\frac{1}{3}t+\frac{1}{2}t)=6$$

$$\frac{6}{1}\left(\frac{1}{3}\right)t+\frac{6}{1}\left(\frac{1}{2}\right)t=6$$

$$\frac{\cancel{6}^{2}}{1}\left(\frac{1}{\cancel{3}_{1}}\right)t + \frac{\cancel{6}^{3}}{1}\left(\frac{1}{\cancel{2}_{1}}\right)t = 6$$

$$2t + 3t = 6$$
$$5t = 6$$
$$\frac{5t}{5} = \frac{6}{5}$$
$$t = 1\frac{1}{5}$$

Answer the question and include the units: They can maintain Mrs. Bilski's lawn in $1\frac{1}{5}$ hours if they work together.

5. **Check the solution.**
   a. Estimation: Does the answer seem reasonable? If there were two of Martin, they could both maintain the lawn in $1\frac{1}{2}$ hours. If there were two of Nabil, it would take one hour. $1\frac{1}{5}$ is between one and $1\frac{1}{2}$ hours and is a reasonable answer.
   b. Checking the solution in the equation:

$$\frac{1}{3}\left(\frac{6}{5}\right) + \frac{1}{2}\left(\frac{6}{5}\right) = 1$$

$$\frac{1}{\cancel{3}_{1}}\left(\frac{\cancel{6}^{2}}{5}\right) + \frac{1}{\cancel{2}_{1}}\left(\frac{\cancel{6}^{3}}{5}\right) = 1$$

$$\frac{2}{5} + \frac{3}{5} = 1$$
$$\frac{5}{5} = 1$$
$$1 = 1 \checkmark$$

Example 11: Two inlet pipes are filling the city pool. One inlet pipe can fill the pool in 15 hours. The other inlet pipe can fill the pool in 12 hours. How long will it take for both inlet pipes together to fill the city pool?

1. **Read the problem.**
2. **Identify the variables.**

   t, the time it takes to fill the pool with both inlet pipes

3. Write the equation.

Find the portion of the pool will be filled by the first pipe in one hour. The flow rate of the first inlet pipe is $\frac{1}{15}$ of the pool per hour. The flow rate of the second inlet pipe is $\frac{1}{12}$ of the pool per hour. The portions filled by each pipe must add up to the whole job, 1.

$$\frac{1}{15}t + \frac{1}{12}t = 1$$

4. Solve the equation.

To clear fractions, multiply every term on both sides of the equation by the least common denominator (LCD). The least common multiple of 15 and 12 is 60.

$$60\left(\frac{1}{15}t + \frac{1}{12}t\right) = 60(1)$$

$$60\left(\frac{1}{15}t + \frac{1}{12}t\right) = 60$$

$$\frac{60}{1}\left(\frac{1}{15}\right)t + \frac{60}{1}\left(\frac{1}{12}\right)t = 60$$

$$\frac{\overset{4}{\cancel{60}}}{1}\left(\frac{1}{\underset{1}{\cancel{15}}}\right)t + \frac{\overset{5}{\cancel{60}}}{1}\left(\frac{1}{\underset{1}{\cancel{12}}}\right)t = 60$$

$$4t + 5t = 60$$

$$9t = 60$$

$$\frac{9t}{9} = \frac{60}{9}$$

$$t = 6\frac{2}{3}$$

Answer the question and include the units: The two inlet pipes can fill the city pool in $6\frac{2}{3}$ hours.

5. Check the solution.

a. Estimation: Does the answer seem reasonable? If both pipes flowed at the rate of one pool in 15 hours, the pool would be filled in $7\frac{1}{2}$ hours. If both pipes flowed at the rate of one pool in 12 hours, the pool would be filled in 6 hours. $6\frac{2}{3}$ is between six and $7\frac{1}{2}$ hours and is a reasonable answer.

b. Checking the solution in the equation:

$$\frac{1}{15}\left(\frac{60}{9}\right)+\frac{1}{12}\left(\frac{60}{9}\right)=1$$

$$\frac{1}{\cancel{15}_{1}}\left(\frac{\cancel{60}^{4}}{9}\right)+\frac{1}{\cancel{12}_{1}}\left(\frac{\cancel{60}^{5}}{9}\right)=1$$

$$\frac{4}{9}+\frac{5}{9}=1$$

$$\frac{9}{9}=1$$

$$1=1\ \checkmark$$

Chapter Checkout

Q&A

Solve the following problems. Be sure to answer the question and include the units.

1. Find three consecutive integers such that the difference between three times the third and the first is equal to sum of the second and four.
2. Find two consecutive integers such that seven times the first integer minus 23 equals the second integer.
3. Find two consecutive odd integers such that the difference between three times the first odd integer and the second odd integer is 16.
4. Find three consecutive even integers such that the sum of the first and third is three times the second even integer minus 4.
5. Find three consecutive odd integers such that the first plus the second is equal to negative four times the third.
6. The life expectancy of a person living in Wyoming is 1.5 years longer than a person living in New York state. If the life expectancy of a person living in Wyoming is 76.2 years, what is the life expectancy of a person living in New York state?
7. Twinkle Toes is an apropos name for a cat with more toes than the average cat. The average cat has 5 toes on each of the front paws and 4 toes on each back paw for a total of 18 toes. The average cat has 7 toes fewer than Twinkle Toes. How many toes does Twinkle Toes have?

8. Leonardo da Vinci began his career as an artist's apprentice at age 14. Thirty-seven years later, he began painting the *Mona Lisa.* How old was Leonardo da Vinci when he began painting the *Mona Lisa?*

9. The primary language spoken by 341 million people is English. Chinese is the primary language spoken by 190 million more than three times the number of people speaking English. How many millions of people speak Chinese?

10. The average height of an adult cocker spaniel is fifteen inches. A Great Dane stands twice the height of a cocker spaniel, plus two inches. Find the average height of an adult Great Dane.

11. A machine shop in Topeka has a new machine that can tool enough parts to fill the regular monthly order for the local dealer in only 3 hours. The old machine took 6 hours to fill the same order. How long will it take to tool the parts if both machines are used?

12. Stephanie's catering service provides place settings for each of the wedding receptions they cater. Lua can wash 250 place settings in 4 hours. Kinnie can wash 250 place settings in 3 hours. If they work together, how long will it take them to wash 250 place settings?

Answers: 1. The three consecutive integers are –1, 0, and 1. **2.** The two consecutive integers are 4 and 5. **3.** The two consecutive odd integers are 9 and 11. **4.** The three consecutive even integers are 2, 4, and 6. **5.** The three consecutive odd integers are –3, –1, and 1. **6.** The life expectancy of a person living in New York state is 74.7 years. **7.** Twinkle toes has 25 toes. **8.** Leonardo da Vinci was 51 years old when he began painting the *Mona Lisa.* **9.** There are 1,213 million people who speak Chinese. **10.** An adult Great Dane stands 32 inches tall. **11.** It will take 2 hours if both machines are used. **12.** It will take them $1\frac{5}{7}$ hours to wash 250 place settings.

Chapter 9

SYSTEMS OF EQUATIONS USING THE BOARD METHOD

Chapter Check-In

- ❑ Money problems
- ❑ Investment problems
- ❑ Mixture problems
- ❑ Distance problems

In this chapter, you learn to recognize word problems with two totals or a combination of a total and a sentence that translates into an equation, and you solve them using the board method discussed in Chapter 7. At the end of this chapter, you also solve those infamous train problems (such as, "Two trains leave the station at the same time going in opposite directions").

Money Problems

A common type of word problem is one concerning money. Many money problems can be represented with one or two boards. The boards make the problem easy to visualize and, therefore, easy to solve.

Example 1: The football game tickets cost $5 per adult and $2 per child. A total of $908 was collected at the ticket booth. Two hundred twenty tickets were sold that night. How many of each type of ticket were sold?

If the problem gives you at least one total, a board can be used to solve the problem. Example 1 has two totals. One is $908, and the other is 220 tickets. Because this problem has totals, use the six steps for solving word problems introduced in Chapter 7.

1. **Read the problem.**
2. **Draw the diagram(s).**

   When you find a word problem containing two totals, draw two boards. If one of the totals gives you a total *number* of items, draw a board for that total first. Then draw a board that shows the *value* of those items. In this case, the first board will show 220 tickets, while the second board will show $908.
3. **Label the diagram(s).**

   a. Board one:

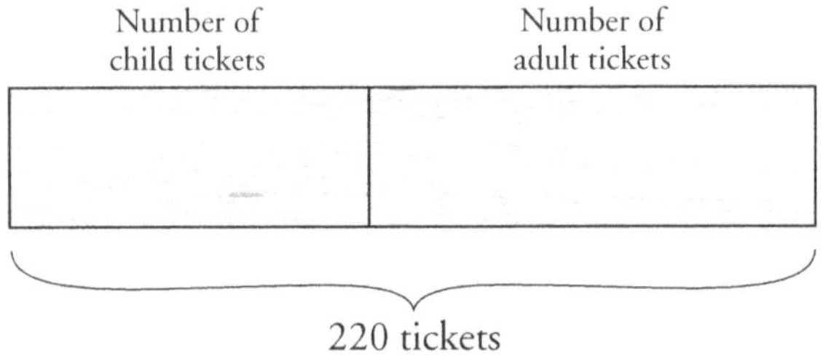

You can only guess which number goes in the smaller piece of the board because no information is given about the relationship of the two unknowns. Label the smaller piece as your variable, and then define the other in terms of that variable.

Note: When using tried-and-true methods to solve word problems (see Chapters 8 & 10), the second step is to identify the unknown(s) and write a brief description. When using the board method, you identify the unknown(s) when you label the board; that is, the description of the variable is written directly above the board diagram.

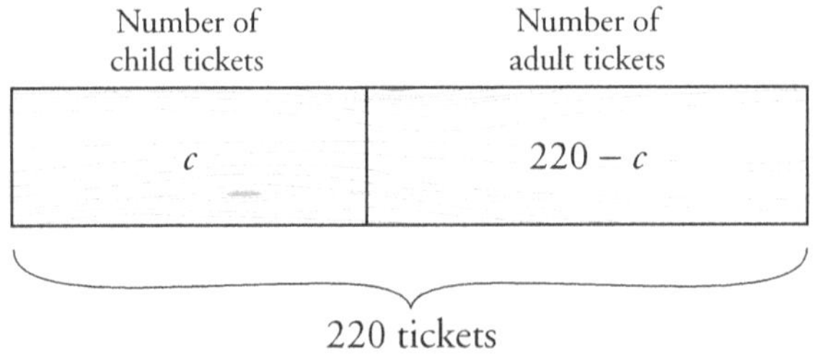

You may be asking yourself where "220 – *c*" came from. If you are not sure, use the following technique: When something is difficult to understand using a variable (which is **abstract**), use numbers (which are **concrete**) to find the pattern (or operation) needed to solve the problem. For example: If they sold 60 child tickets, how would you find the number of adult tickets? 220 – 60 = 160 adult tickets. Thus, the number of adult tickets is found by taking the total number of tickets minus the number of child tickets, hence 220 – *c*.

If you try to write an equation from this single board, the equation becomes

$$c + (220 - c) = 220$$

Simplifying the left side of the equation yields

$$220 = 220$$

This is always true, no matter what value the variable *c* has. Therefore, you cannot write a useful equation from this board alone.

b. Board two:

Draw a board to represent the total value of the tickets. To find the value of the children's tickets, multiply the value of one ticket times the number of child tickets sold, hence 2*c* is the total value of the child tickets sold. Similarly, you can find the total value of the adult tickets.

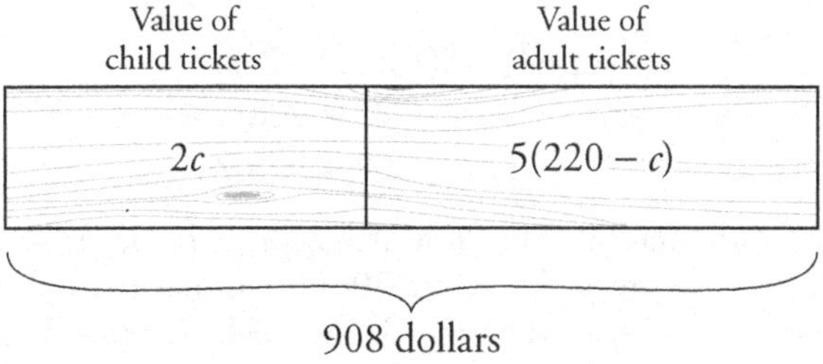

Note: The first board shows that the children's tickets plus the adult tickets equal the total number of tickets. The second board shows that the value of child tickets plus the value of adult tickets equal the total value of tickets. Be sure to add apples to apples. In other words, when finding a total value of tickets, add the value of the tickets, not the number of tickets.

4. Write an equation.

$$2c + 5\,(220 - c) = 908$$

5. Solve the equation.

$$2c + 5(220 - c) = 908$$
$$2c + 1{,}100 - 5c = 908$$
$$-3c + 1{,}100 = 908$$
$$-3c + 1{,}100 - 1{,}100 = 908 - 1{,}100$$
$$-3c = -192$$
$$\frac{-3c}{-3} = \frac{-192}{-3}$$
$$c = 64$$

The variable *c* represents the number of children's tickets. Looking at the board representing the total number of tickets, you can find the number of adult tickets by subtracting: 220 – 64.

Answer the question and include the units: 64 children's tickets and 156 adult tickets were sold at the football game.

6. Check the solution(s).

a. Number of items:

$$64 + 156 = 220$$
$$220 = 220 \checkmark$$

b. Value of items:

$$2(64) + 5(156) = 908$$
$$128 + 780 = 908$$
$$908 = 908 \checkmark$$

Example 2: Sandra has 62 coins in Martha's piggy bank. All the coins are dimes, nickels, or pennies. There are 20 more dimes than pennies. The total value of the money in the piggy bank is $4.27. How many of each type of coin does Sandra have?

1. Read the problem, making helpful markings.

There are 20 more dimes [than] pennies.

See Chapter 1 for more on turnaround words.

2. Draw the diagram(s).

The problem gives you two totals; therefore, draw two boards. Draw the first board for the number of items (62 coins); draw the second board to the total value ($4.27).

3. Label the diagram(s).

a. Board one:

Number of pennies	Number of nickels	Number of dimes
p		$p + 20$

62 coins

Did you estimate that the number of dimes is the largest number and the number of pennies is the smallest? That is correct, but how do you find the number of nickels when that relationship is not given. Use concrete numbers to find out the correct relationship. For example: If Sandra has 10 pennies, how many dimes did she have? She has 20 more dimes than pennies: 10 + 20 = 30 dimes. How would you then find the number of nickels? The total number of nickels is the total number of coins minus all coins that are not nickels. Therefore, you add the number of pennies and the number of dimes:

$$10 + 30 = 40$$

and subtract the answer from 62:

$$62 - 40 = 22 \text{ nickels}$$

Now, you can do this abstractly, with variables.

Number of pennies	Number of nickels	Number of dimes
p	$62 - [p + (p + 20)]$	$p + 20$

62 coins

Simplify:

Number of pennies	Number of nickels	Number of dimes
p	$42 - 2p$	$p + 20$

62 coins

This equation alone, however, cannot be solved for p.

$$\begin{aligned} p + (42 - 2p) + (p + 20) &= 62 \\ p + -2p + p + 42 + 20 &= 62 \\ 42 + 20 &= 62 \\ 62 &= 62 \end{aligned}$$

This answer is true no matter what is the value of p. So, this board is helpful only in establishing the relationship of the number of nickels and dimes to the number of pennies.

b. Board two:

You need to sketch the board for the total value. As in the Example 1, find the total value by taking the value of one item and multiplying it by the number of items.

Total value of pennies	Total value of nickels	Total value of dimes
$0.01p$	$0.05(42 - 2p)$	$0.10(p + 20)$

\$4.27

4. Write the equation.

$$0.01p + 0.05(42 - 2p) + 0.10\,(p + 20) = 4.27$$

5. Solve the equation.

$$0.01p + 0.05(42 - 2p) + 0.10(p + 20) = 4.27$$

$$0.01p + 2.10 - 0.10p + 0.10p + 2.00 = 4.27$$

Clear the decimals by multiplying each side of the equation by 100.

$$1p + 210 - 10p + 10p + 200 = 427$$
$$1p + 410 = 427$$
$$1p + 410 - 410 = 427 - 410$$
$$p = 17$$

Answer the question and include the units: Sandra has 17 pennies, 37 dimes, and 8 nickels in Martha's piggy bank.

These numbers are found by taking the number of pennies, 17, and looking at the first board to find the number of nickels and dimes.

6. **Check the solutions.**

a. Number of items:

\# of pennies + # of dimes + # of nickels = total # of coins

$$17 + 37 + 8 = 62$$
$$54 + 8 = 62$$
$$62 = 62 \checkmark$$

b. Value of items:

$$0.01(17) + 0.05(8) + 0.10(37) = 4.27$$
$$0.17 + 0.40 + 3.70 = 4.27$$
$$0.57 + 3.70 = 4.27$$
$$4.27 = 4.27 \checkmark$$

Investment Problems

You may be tempted to work investment problems like the money problems in the preceding section. While many of the same methods are used in investment problems, you do find a subtle difference between investment problems and money problems, as you see in Examples 3 and 4.

Example 3: A total of $10,550 is invested. Part of the money earns an annual percentage rate (APR) of 5%, and the rest of the money is invested at an APR of 7%. If the total simple interest earned after one year is $648.50, how much was invested at each rate?

1. **Read the problem.**

   The problem gives two totals: a total investment of \$10,550 and total interest of \$648.50.

   Remember: Because the money was invested at a simple interest rate, to find the interest earned in one year, you multiply the interest rate times the amount invested at that rate.

2. **Draw the diagram(s).**

   Although \$648.50 is a smaller number, it is the total value of the interest, whereas \$10,550 is the total number of dollars invested. Use \$10,500 for the first board because it represents the total number of items. Use \$648.50, the total value, for the second board.

3. **Label the diagram(s).**

   a. Board one:

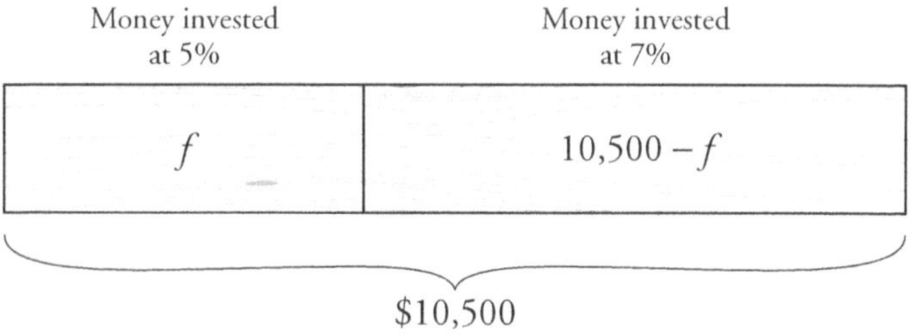

   Note: The variable, *f*, is assigned to the smaller piece of the board. The relationship between the two amounts invested is found by taking the total amount invested and subtracting the amount invested at 5%.

   b. Board two:

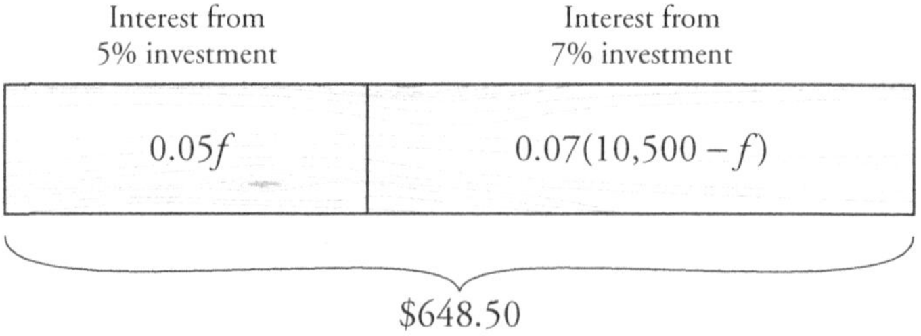

   Find the value of the interest by multiplying the number of items by the value of one item. In other words, multiply the number of dollars invested at 5%, *f*, times the value of interest for one dollar, 0.05.

4. **Write the equation.**

$$0.05f + 0.07(10{,}500 - f) = 648.50$$

5. **Solve the equation.**

$$0.05f + 0.07(10{,}500 - f) = 648.50$$

$$0.05f + 738.50 - 0.07f = 648.50$$

Clear the decimals by multiplying both sides of the equation by 100.

$$5f + 73{,}850 - 7f = 64{,}850$$
$$-2f + 73{,}850 = 64{,}850$$
$$-2f + 73{,}850 - 73{,}850 = 64{,}850 - 73{,}850$$
$$-2f = -9{,}000$$
$$\frac{-2f}{-2} = \frac{-9{,}000}{-2}$$
$$f = 4{,}500$$

Answer the question and include the units: \$4,500 was invested at 5%, and \$6,050 was invested at 7%.

6. **Check the solution(s).**

a. Number of items:

$$4{,}500 + 6{,}050 = 10{,}550$$
$$10{,}550 = 10{,}550 \checkmark$$

b. Value of items:

$$0.05(4{,}500) + 0.07(6{,}050) = 648.50$$
$$225 + 423.50 = 648.50$$
$$648.50 = 648.50 \checkmark$$

Example 4: An investment banker invests \$7,600 for a client. The client is conservative and wants most of the money invested in an account with a fixed simple interest rate of 6%. The client takes the advice of the investment banker and invests the rest in a higher-risk account, a technology fund. After a year, the client has earned \$567 in interest. The technology fund averaged a 9% simple interest return. How much did the client invest in each account?

1. **Read the problem.**
2. **Draw the diagram(s).**

   Notice that \$7,600 goes on the first board because it represents the total number of dollars invested. The total value (\$567, the interest earned) is represented on the second board.
3. **Label the diagram(s).**

   a. Board one:

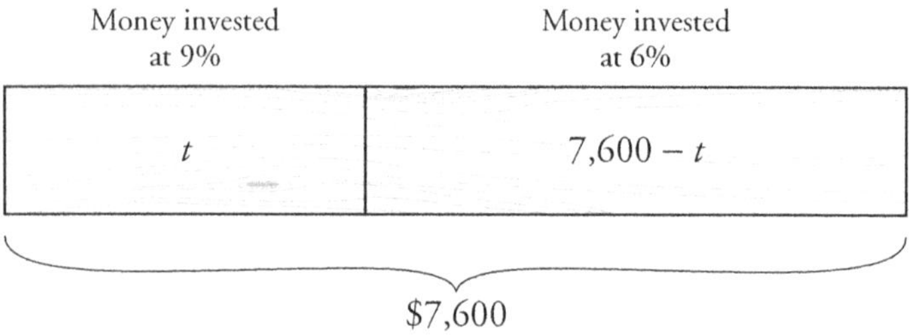

Note: The money invested at 6% is on the larger portion of the board because the client wanted most of his money in the account with a fixed simple interest rate of 6%.

b. Board two:

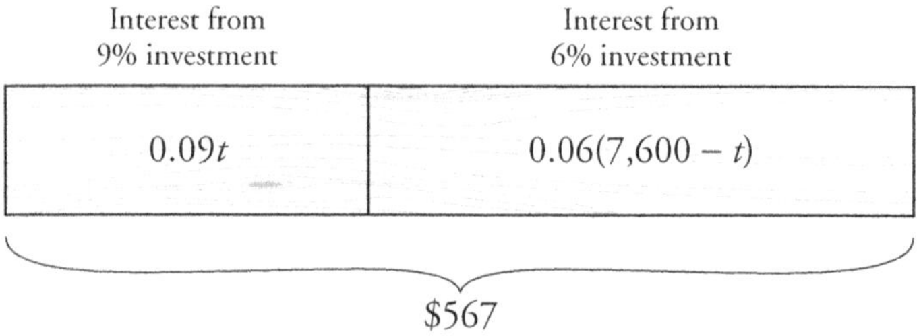

The value of interest from each investment is found by multiplying the value of interest on one dollar by the number of dollars invested.

4. **Write the equation.**

$$0.09t + 0.06(7{,}600 - t) = 567$$

5. **Solve the equation.**

$$\begin{aligned} 0.09t + 0.06(7{,}600 - t) &= 567 \\ 0.09t + 456 - 0.06t &= 567 \end{aligned}$$

Multiply every term on both sides of the equation by 100 to clear the decimals.

$$9t + 45{,}600 - 6t = 56{,}700$$
$$3t + 45{,}600 = 56{,}700$$
$$3t + 45{,}600 - 45{,}600 = 56{,}700 - 45{,}600$$
$$3t = 11{,}100$$
$$\frac{3t}{3} = \frac{11{,}100}{3}$$
$$t = 3{,}700$$

Answer the question and include the units: \$3,700 was invested in the technology fund that yielded 9% simple interest; \$3,900 was invested in the fixed account that yielded 6% simple interest.

6. **Check the solution(s).**

   a. Number of items:

$$\$3{,}700 + \$3{,}900 = \$7{,}600$$
$$\$7{,}600 = \$7{,}600 \checkmark$$

   b. Value of items (interest):

$$0.06(3{,}900) + 0.09(3{,}700) = 567$$
$$234 + 333 = 567$$
$$567 = 567 \checkmark$$

Mixture Problems

Mixture problems follow roughly the same pattern as money problems and investment problems. These problems often give you one total, and the information needed to find another total by multiplying the number of items by the value of each item.

Example 5: In a chemistry lab, students are told to mix a 26% hydrochloric-acid solution (Solution A) with a 6% hydrochloric-acid solution (Solution B) to produce a 45-milliliter mixture of 18% hydrochloric acid. How many milliliters of each solution must be used to produce the correct mixture?

1. **Read the problem.**
2. **Draw the diagram(s).**

   You are given a total number of milliliters and the total percent value of the final solution. Have the first board show the total number of milliliters of solution; the second board can show the value of the acid in the final solution.

3. Label the diagram(s).

a. Board one:

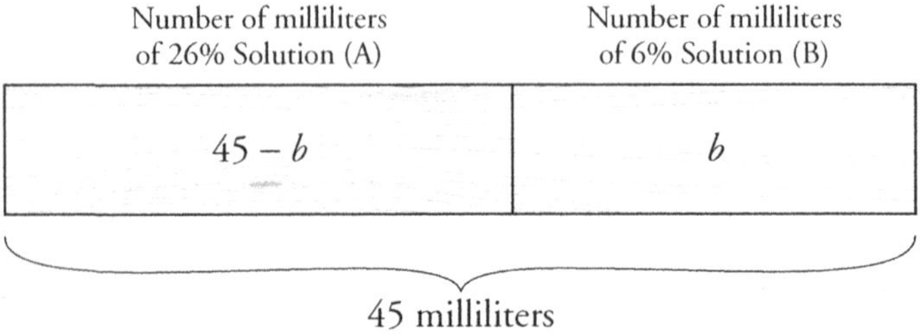

Note: The larger portion of the board is labeled with the number of milliliters of 26% solution because 18% (the final percentage) is closer to 26% than 6% is.

b. Board two:

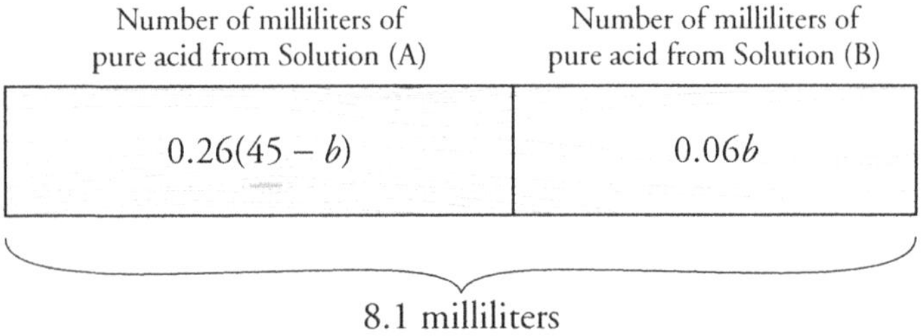

The total value of the acid in the final solution is found by multiplying the value of acid of one milliliter by the total number of milliliters. In other words, multiply 0.18 by 45. The total number of milliliters of pure acid in the final mixture is 8.1 milliliters. Add milliliters of acid in Solution A to milliliters of acid in Solution B to get the total amount of acid in the final solution.

4. Write the equation.

$$0.26\,(45 - b) + 0.06\,(b) = 8.1$$

5. Solve the equation.

$$0.26(45 - b) + 0.06b = 8.1$$
$$11.70 - 0.26b + 0.06b = 8.1$$

Multiply every term on both sides of the equation by 100.

$$1{,}170 - 26b + 6b = 810$$
$$-20b + 1{,}170 = 810$$
$$-20b + 1{,}170 - 1{,}170 = 810 - 1{,}170$$
$$-20b = -360$$
$$\frac{-20b}{-20} = \frac{-360}{-20}$$
$$b = 18$$

Answer the question and include the units: 18 milliliters are needed of the 6% hydrochloric-acid solution (Solution B); 27 milliliters are needed of the 26% hydrochloric-acid solution (Solution A).

6. **Check the solution(s).**

   a. Number of milliliters (acid and water):

$$18 + 27 = 45$$
$$45 = 45 \checkmark$$

   b. Value of items (acid):

$$0.26(27) + 0.06(18) = 0.18(45)$$
$$1.08 + 7.02 = 8.1$$
$$8.1 = 8.1 \checkmark$$

Example 6: Nequoia has created her own flavor of coffee by grinding two types of flavored coffee beans together. She mixes amaretto-flavored beans and chocolate-flavored beans to form a five-pound blend. Amaretto-flavored beans cost $8.95 per pound, while chocolate-flavored beans cost only $7.95 per pound. The five-pound mixture costs $41.75. How many pounds of each type of bean does she use for her special blend?

1. **Read the problem.**

   The problem gives you two totals: five pounds and $41.75.

2. **Draw the diagram(s).**

   Have the first board show the total number of pounds of coffee (five). Use the total value of the coffee ($41.75) for the second board.

3. Label the diagram(s).

a. Board one:

Number of pounds of amaretto beans	Number of pounds of chocolate beans
a	$5 - a$

5 pounds

b. Board two:

Value of amaretto beans	Value of chocolate beans
$8.95a$	$7.95(5 - a)$

\$41.75

4. Write the equation.

$$8.95\,(a) + 7.95\,(5 - a) = 41.75$$

5. Solve the equation.

$$8.95a + 7.95(5 - a) = 41.75$$
$$8.95a + 39.75 - 7.95a = 41.75$$

Multiply every term on both sides of the equation by 100 to clear the decimals.

$$895a + 3{,}975 - 795a = 4{,}175$$
$$100a + 3{,}975 = 4{,}175$$
$$100a + 3{,}975 - 3{,}975 = 4{,}175 - 3{,}975$$
$$100a = 200$$
$$\frac{100a}{100} = \frac{200}{100}$$
$$a = 2$$

Answer the question and include the units: Nequoia's special blend of flavored coffees uses two pounds of amaretto-flavored beans and three pounds of chocolate-flavored beans.

6. **Check the solution(s).**

   a. Number of pounds of coffee:

$$\begin{aligned} 2 + 3 &= 5 \\ 5 &= 5 \checkmark \end{aligned}$$

   b. Value of items:

$$\begin{aligned} 8.95(2) + 7.95(3) &= 41.75 \\ 17.90 + 7.95(3) &= 41.75 \\ 17.90 + 23.85 &= 41.75 \\ 41.75 &= 41.75 \checkmark \end{aligned}$$

Distance Problems

In order to be able to work distance problems, you need to know the distance, d, that an object travels. You can find this distance by multiplying the rate, r, by the number of hours, t, the object travels at that rate:

$$d = r \times t \text{ or } d = rt$$

If a distance problem gives you a total, the board method can help you diagram the problem. If you are not given a total, see Chapter 10.

Example 7: Two trains leave the station at the same time going in opposite directions. After three hours, the trains are 408 miles apart. The difference in their average speeds is only 4 miles per hour. Find the speed of both trains.

1. **Read the problem.**
2. **Draw the diagram(s).**

   This problem has only one total and needs only one board. Draw a board with the total number of miles the trains are apart.
3. **Label the diagram(s).**

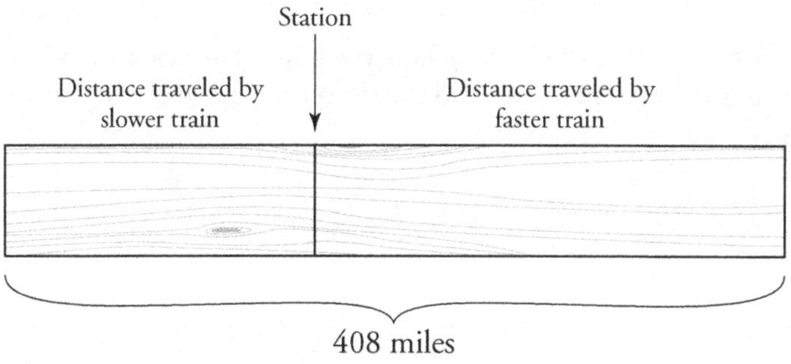

Use the formula $d = rt$ to find the distance traveled by each train. Both trains are traveling for three hours, so $t = 3$ for both trains. The rate of each train is unknown. There is a difference of four miles per hour in their average speed. If one train travels at a rate of speed, r, the speed of the other train can be written $r - 4$.

Remember: The train traveling at $r - 4$ is the slower train.

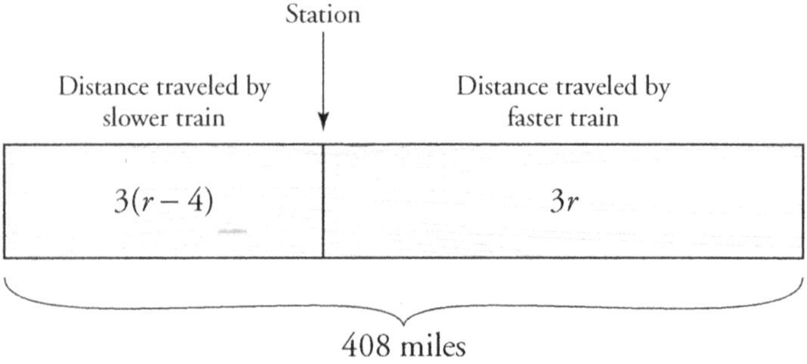

The value (in this case, the distance) is found by multiplying the value of one hour by the number of hours.

4. **Write the equation.**

$$3(r - 4) + 3(r) = 408$$

5. **Solve the equation.**

$$3(r - 4) + 3r = 408$$
$$3r - 12 + 3r = 408$$
$$6r - 12 = 408$$
$$6r - 12 + 12 = 408 + 12$$
$$6r = 420$$
$$\frac{6r}{6} = \frac{420}{6}$$
$$r = 70$$

Answer the question and include the units: The faster train is averaging 70 miles per hour, and the slower train is averaging 66 miles per hour.

6. **Check the solution(s).**

   Value of items:

$$70(3) + 66(3) = 408$$
$$210 + 66(3) = 408$$
$$210 + 198 = 408$$
$$408 = 408 \checkmark$$

Example 8: Michael and Brandon kayak together often and have found that they average the same speed in still water. After two hours of kayaking from the same beginning point on the river, they are thirty-two miles apart. Michael travels upstream, while Brandon travels twenty-two miles downstream. Find the speed of the current of the river and the speed of Michael and Brandon in still water.

1. **Read the problem.**
2. **Draw the diagram(s).**

   Draw one board with the total number of miles the kayaks are apart.
3. **Label the diagram(s).**

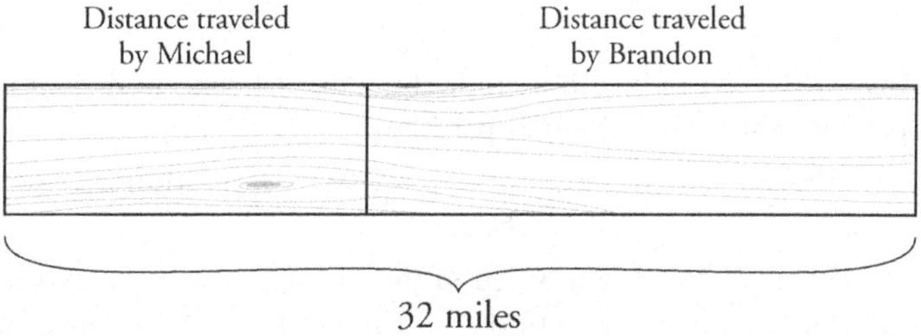

The shorter piece of the board is labeled Michael because he is traveling against the current and, therefore, is not able to travel as far in two hours. Brandon is traveling with the current and is going farther than he would in still water for two hours.

Use the formula $d = rt$ to find the distance traveled by each kayak. Both kayaks are traveling for two hours, so $t = 2$. Michael's rate is slowed by the current, so his rate can be represented by $r - c$. Because Brandon's rate is increased by the current, his rate is $r + c$.

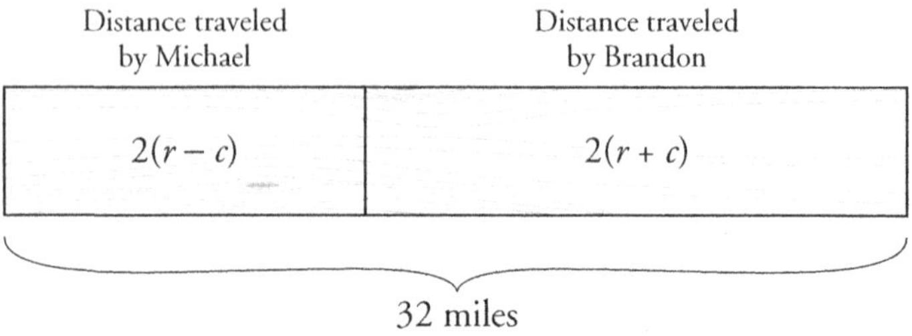

4. Write the equation.

$$2(r - c) + 2(r + c) = 32$$

5. Solve the equation.

$$2(r - c) + 2(r + c) = 32$$
$$2r - 2c + 2r + 2c = 32$$
$$4r = 32$$
$$\frac{4r}{4} = \frac{32}{4}$$
$$r = 8$$

Answer the question and include the units: Michael and Brandon's speed in still water is 8 miles per hour.

Does this answer the entire question? No, the question also asks for the speed of the current of the river.

The distance Brandon traveled in the two hours is not necessary to find *r*, but is necessary to find *c*. Use the formula $d = rt$. Substitute the known values and solve for the speed of the current.

$$d = rt$$
$$22 = (r + c)2$$
$$22 = (8 + c)2$$

$$22 = (8 + c)2$$
$$22 = 16 + 2c$$
$$22 - 16 = 16 - 16 + 2c$$
$$6 = 2c$$
$$\frac{6}{2} = \frac{2c}{2}$$
$$3 = c$$
$$c = 3$$

Answer the question and include the units: Michael and Brandon's speed in still water is 8 miles per hour. The speed of the current of the river is three miles per hour.

6. Check the solution(s).

a. Equation one:

$$2(8-3)+2(8+3)=32$$
$$2(5)+2(8+3)=32$$
$$2(5)+2(11)=32$$
$$10+2(11)=32$$
$$10+22=32$$
$$32=32\ ✓$$

b. Equation two:

$$2(8+3)=22$$
$$2(11)=22$$
$$22=22\ ✓$$

Chapter Checkout

Q&A

Solve the following problems. Be sure to answer the question and include the units.

1. A community theater group is performing *Arsenic and Old Lace.* A patron can come at 8:00 for the play or at 7:00 for dinner and the play. The dinner theater tickets cost $25 each. The play alone costs $10 per ticket. Opening night played to a full house, 125 seats, and brought in $2,345. How many of each type of ticket were sold?
2. The Hamilton High School drill team held a garage sale to raise money for new uniforms. The fundraiser brought in $2,007.75 in twenties, tens, fives, ones, and quarters. There are fourteen times as many ones as five dollar bills, six fewer tens than fives, ten times as many twenties as fives, and three times as many quarters as five dollar bills. How many of each does the drill team have?
3. Lisa received an inheritance of $33,000. She wants to invest in a fixed savings account and in supplies necessary to set up a home business. The fixed account earned 6.75% in simple interest, and her new business made an 8% profit. Her total profit from the two investments for one year was $2,402.50. How much did Lisa invest in each?

4. A natural fertilizer is made by mixing household compost and rabbit manure. Stan wants his fertilizer to have 75% rabbit manure. He still has some of last year's fertilizer, which was made up of 60% rabbit manure. How many gallons of pure 100% rabbit manure and last year's fertilizer must be mixed together to yield six gallons of a fertilizer having 75% manure?

5. In chemistry lab, Ben is asked to make a 70-milliliter solution of 15% sulfuric acid. He is given two solutions: Solution A is a 10% sulfuric acid solution, and solution B is a 24% sulfuric acid solution. How many milliliters of each solution must Ben use to produce the correct mixture?

6. A train leaves the station at 9 a.m. traveling east. Another train also leaves the station at 9 a.m. traveling west. At noon, they are 465 miles apart. The train traveling east is averaging 25 fewer miles per hour than the train traveling west. Find the speed of each train.

7. Dennis can swim freestyle at a speed of 4 mph. His average speed running is 9 mph, and he can bicycle 22 mph. At a triathlon competition in Albany, he completed the 68 mile race in 4.5 hours. He spent the same time swimming as running. How long was each part of the race?

Answers: 1. Seventy-three dinner theater tickets were sold, and fifty-two patrons bought tickets for the play without dinner. **2.** The drill team has 90 twenty dollar bills, 3 tens, 9 fives, 126 ones, and 27 quarters. **3.** Lisa invested $19,000 in the fixed account and $14,000 in the new business. **4.** The new fertilizer needs to be mixed from 2.25 gallons of pure rabbit manure and 3.75 gallons of last year's fertilizer. **5.** Ben must mix 45 ml of solution A with 25 ml of solution B. **6.** The train traveling east is traveling at a speed of 65 mph. The westbound train is traveling 90 mph. **7.** The swimming portion of the race was 4 miles, the running part was 9 miles, and the bicycling part was 55 miles.

Chapter 10

SYSTEMS OF EQUATIONS USING TRIED-AND-TRUE METHODS

Chapter Check-In

- ❑ Money problems
- ❑ Investment problems
- ❑ Mixture problems
- ❑ Distance problems

Not all word problems with multiple equations, also called systems of equations, can be solved using the board method described in Chapter 9. When a problem has two unknowns, you can write two equations and solve the system of equations by the **substitution method** or the **elimination method.** These tried-and-true methods for solving systems of equations are presented in this chapter. The examples at the beginning of this chapter are the same as the examples in Chapter 9. This is to introduce the tried-and-true methods of substitution and elimination with problems that you are already familiar with. Later in this chapter examples are introduced that cannot be solved by the board method and must be solved by tried-and-true methods.

Money Problems

Examples 1 and 2 in this chapter are also Examples 1 and 2 in Chapter 9, where the board method is used to solve them. Here, tried-and-true methods are used: The elimination method is used in Example 1, and the substitution method is used in Example 2.

Example 1: The football game tickets cost \$5 per adult and \$2 per child. A total of \$908 was collected at the ticket booth. Two hundred and twenty tickets were sold that night. How many of each type of ticket were sold?

1. **Read the problem.**
2. **Identify the variables.**

a, the number of adult tickets sold
c, the number of children's tickets sold

3. **Translate the equations.**

$$a + c = 220$$
$$5a + 2c = 908$$

As you translate the equations, remember to add apples to apples. All units should be the same for every term in the equation. In the first equation, the number of children's tickets is added to the number of adult tickets to get a total number of tickets. In the second equation, every term in the equation is measured in dollars.

4. **Solve the system of equations.**

Use the **elimination method,** which is sometimes called the **addition method.** One (or both) of the equations needs to be modified so that when the two equations are added, one of the variables is eliminated. In this example, the variable c is easy to eliminate by multiplying both sides of the first equation by negative two, and then adding the two equations.

$$-2(a + c) = -2(220)$$

$$-2(a + b) = -440$$

$$-2a + -2c = -440$$

Add the modified version of the first equation to the second equation and solve for the unknown variable.

$$\begin{array}{r} -2a + -2c = -440 \\ +5a + 2c = 908 \\ \hline 3a = 468 \end{array}$$

$$\frac{3a}{3} = \frac{468}{3}$$

$$a = 156$$

The variable a represents the number of adult tickets (as noted in Step 2.)

Use either of the original equations to solve for c.

$$\begin{aligned} a + c &= 220 \\ 156 + c &= 220 \\ 156 + c - 156 &= 220 - 156 \\ c &= 64 \end{aligned}$$

Answer the question and include the units: 64 children's tickets and 156 adult tickets were sold at the football game.

5. **Check the solutions.**

   a. First equation:

$$\begin{aligned} 64 + 156 &= 220 \\ 220 &= 220 \checkmark \end{aligned}$$

   b. Second equation:

$$\begin{aligned} 2(64) + 5(156) &= 908 \\ 128 + 780 &= 908 \\ 908 &= 908 \checkmark \end{aligned}$$

Example 2 is a problem that was solved using the board method in Chapter 9. The tried-and-true methods will be used in this chapter to illustrate the substitution method of solving a system of equations.

Example 2: Sandra has 62 coins in Martha's piggy bank. All of the coins are dimes, nickels, or pennies. There are 20 more dimes than pennies. The total value of the money in the piggy bank is \$4.27. How many of each type of coin does Sandra have?

1. **Read the problem, making helpful markings.**

   There are 20 more dimes [than] pennies.

2. **Identify the variables.**

   d, the number of dimes
   n, the number of nickels
   p, the number of pennies

   This problem has three unknowns, so three equations are needed.

3. **Translate the equations.**

   Reread the problem, translating one sentence at a time into an equation, if possible. **Note:** The second sentence does not translate into an equation.

   a. Equation one:

   The first sentence translates as follows:

$$d + n + p = 62$$

   b. Equation two:

   When translating the third sentence, use your estimation skills. Notice there are more dimes than pennies, so the number of dimes, *d,* must be more than the number of pennies, *p*.

$$d = p + 20$$

   c. Equation three:

   The fourth sentence translates as follows:

$$0.01p + 0.05n + 0.10d = 4.27$$

4. **Solve the system of equations.**

   With three equations and three unknowns, you can use the substitution method to solve the system of equations.

   The second equation is solved for d in terms of p; $(p + 20)$ can be substituted in both of the other equations.

$$(p + 20) + n + p = 62$$
$$0.01p + 0.05n + 0.10\,(p + 20) = 4.27$$

   Now you have two equations with two unknowns. Solve the first equation for n.

$$2p + 20 + n = 62$$
$$2p + 20 + n - 20 = 62 - 20$$
$$2p + n = 42$$
$$2p + n - 2p = 42 - 2p$$
$$n = 42 - 2p$$

   Replace n with $(42 - 2p)$ in the other equation.

$$0.01p + 0.05(42 - 2p) + 0.10(p + 20) = 4.27$$

Solve the single equation with one unknown for p.

$$0.01p + 0.05(42 - 2p) + 0.10(p + 20) = 4.27$$
$$0.01p + 2.10 - 0.10p + 0.10p + 2.00 = 4.27$$

Clear the decimals by multiplying each side of the equation by 100.

$$1p + 210 - 10p + 10p + 200 = 427$$
$$1p + 410 = 427$$
$$1p + 410 - 410 = 427 - 410$$
$$p = 17$$

Use the equation $d = p + 20$ to solve for the number of dimes.

$$d = 17 + 20$$
$$d = 37$$

Use the equation $n = 42 - 2p$ to solve for the number of nickels.

$$n = 42 - 2(17)$$
$$n = 42 - 34$$
$$n = 8$$

Answer the question and include the units: Sandra has 17 pennies, 37 dimes, and 8 nickels in Martha's piggy bank.

5. **Check the solutions.**

   a. Equation one:

$$d + n + p = 62$$
$$37 + 8 + 17 = 62$$
$$45 + 17 = 62$$
$$62 = 62 \checkmark$$

   b. Equation two:

$$d = p + 20$$
$$37 = 17 + 20$$
$$37 = 37 \checkmark$$

c. Equation three:

$$0.01p + 0.05n + 0.10d = 4.27$$
$$0.01(17) + 0.05(8) + 0.10(37) = 4.27$$
$$0.17 + 0.40 + 3.70 = 4.27$$
$$0.57 + 3.70 = 4.27$$
$$4.27 = 4.27 \checkmark$$

Example 3: Nathan owns a snow-cone truck. He sells snow cones for $1.50 each and candy for $0.50 each. After two hours in the park, Nathan sold 85 items. He sold 43 more snow cones than candies. How much money did Nathan take in at the park?

1. **Read the problem, making helpful markings.**

   He sold 43 more snow cones [than] candies.

2. **Identify the variables.**

   s, the number of snow cones sold
   c, the number of candies sold
   m, the amount of money he took in

   With three unknowns in this problem, three equations are needed.

3. **Translate the equations.**

   Reread the problem, translating one sentence at a time into an equation, if possible. **Note:** The first sentence does not translate into an equation.

   a. Equation one:

   The second sentence translates as follows:

   $$1.50s + 0.50c = m$$

   b. Equation two:

   The third sentence translates as follows:

   $$s + c = 85$$

   c. Equation three:

   The fourth sentence translates as follows:

   $$s = c + 43$$

4. **Solve the system of equations.**

   Because the second equation gives number of snow cones in terms of number of candies, you can use the substitution method to solve the system of equations.

   Substitute $(c + 43)$ for s in equation two.

$$(c + 43) + c = 85$$
$$2c + 43 = 85$$
$$2c + 43 - 43 = 85 - 43$$
$$2c = 42$$
$$\frac{2c}{2} = \frac{42}{2}$$
$$c = 21$$

   Use the equation $s = c + 43$ to solve for the number of snow cones.

$$s = 21 + 43$$
$$s = 64$$

   Substitute the values for s and c into equation one.

$$1.50(64) + 0.50(21) = m$$
$$96.00 + 0.50(21) = m$$
$$96.00 + 10.50 = m$$
$$106.50 = m$$

   Answer the question and include the units: Nathan took in \$106.50 selling snow cones and candies in the park.

5. **Check the solutions.**

   a. Equation one:

$$1.50s + 0.50c = m$$
$$1.50(64) + 0.50(21) = 106.50$$
$$96 + 0.50(21) = 106.50$$
$$96 + 10.50 = 106.50$$
$$106.50 = 106.50 \checkmark$$

   b. Equation two:

$$s + c = 85$$
$$64 + 21 = 85$$
$$85 = 85 \checkmark$$

c. Equation three:

$$\begin{aligned} s &= c + 43 \\ 64 &= 21 + 43 \\ 64 &= 64 \checkmark \end{aligned}$$

Investment Problems

Investment problems involve investing money at a simple interest rate. In order to successfully work investment problems, you will need to use the equation: $I = prt$. I is the interest earned, p is the principal or the amount of money invested, r is the simple interest rate written as a decimal and t is time in years.

Use the substitution method to solve the system of equations indicated by the word problem. Chapter 9 describes how to use the board method to solve investment problems, but tried-and-true methods are useful when there are no totals.

Example 4: An investment banker invests some money for a client. The client is conservative and wants $200 more money invested in an account with a fixed simple interest rate of 6% than the money invested in a technology fund. After a year, the technology fund averaged a 9% simple interest and earned $333. What was the total amount the client invested and how much did the client have invested in each account?

1. **Read the problem.**
2. **Identify the variables.**

   f, the amount of money invested in the fixed account
   t, the amount of money invested in the technology fund
   T, the total amount of money invested

   You have three unknowns in this problem, so you need three equations.

3. **Translate the equations.**

   Reread the problem, translating one sentence at a time into an equation, if possible. The three equations are

   a. Equation one:

$$f = t + 200$$

b. Equation two:

$$0.09t = 333$$

c. Equation three:

You need three equations and the only sentence left in the problem is the question. The only clue for the third equation is that you need to find the total amount the client invested.

$$f + t = T$$

4. Solve the system of equations.

The second equation is easy to solve for t

$$0.09\,t = 333$$

Multiply every term on both sides of the equation by 100 to clear the decimals.

$$9t = 33{,}300$$
$$\frac{9t}{9} = \frac{33,300}{9}$$
$$t = 3{,}700$$

Replace t with 3,700 in equation one.

$$f = 3{,}700 + 200$$
$$f = 3{,}900$$

Substitute the values for f and t into equation three.

$$3{,}900 + 3{,}700 = T$$
$$7{,}600 = T$$

Answer the question and include the units: \$3,700 was invested in the technology fund that yielded 9% simple interest; \$3,900 was invested in the fixed account that yielded 6% simple interest. A total of \$7,600 was invested by the client.

5. Check the solutions.

a. Equation one:

$$f = t + 200$$
$$3{,}900 = 3{,}700 + 200$$
$$3{,}900 = 3{,}900 \checkmark$$

b. Equation two:

$$0.09t = 333$$
$$0.09(3{,}700) = 333$$
$$333 = 333 \checkmark$$

c. Equation three:

$$f + t = T$$
$$3{,}900 + 3{,}700 = 7{,}600$$
$$7{,}600 = 7{,}600 \checkmark$$

Example 5: Marcos invests \$4,600 in a mutual fund and \$3,200 in savings bonds. After one year, the total interest earned from the investments is \$482. The mutual fund performed 2% better than the savings bonds. Find the rates of interest for the mutual fund and the savings bonds.

1. **Read the problem.**
2. **Identify the variables.**

   m, the interest rate of the mutual fund
   b, the interest rate of the savings bonds

   This problem has two unknowns, so two equations are needed.
3. **Translate the equations.**

   Reread the problem, translating one sentence at a time into an equation, if possible. The two equations are

   a. Equation one:

   $$4{,}600m + 3{,}200b = 482$$

   b. Equation two:

   $$m = b + 0.02$$
4. **Solve the system of equations.**

   Equation two is already solved for m. Use the substitution method to solve the system of equations.

   Replace m with $(b + 0.02)$ in equation one.

   $$4{,}600(b + 0.02) + 3{,}200b = 482$$

   $$4{,}600(b + 0.02) + 3{,}200b = 482$$
   $$4{,}600b + 92 + 3{,}200b = 482$$

$$7{,}800b + 92 = 482$$
$$7{,}800b + 92 - 92 = 482 - 92$$
$$7{,}800b = 390$$
$$\frac{7{,}800b}{7{,}800} = \frac{390}{7{,}800}$$
$$b = 0.05$$

Use the equation $m = b + 0.02$ to solve for the interest rate of the mutual fund.

$$m = 0.05 + 0.02$$
$$m = 0.07$$

Answer the question and include the units: The mutual fund yielded 7% interest, while the rate of interest earned by the savings bond was 5%.

5. **Check the solutions.**

   a. Equation one:

$$4{,}600m + 3{,}200b = 482$$
$$4{,}600(0.07) + 3{,}200(0.05) = 482$$
$$322 + 3{,}200(0.05) = 482$$
$$322 + 160 = 482$$
$$482 = 482 \checkmark$$

   b. Equation two:

$$m = b + 0.02$$
$$0.07 = 0.05 + 0.02$$
$$0.07 = 0.07 \checkmark$$

Mixture Problems

A **mixture problem** is a type of word problem that, when translated into a system of equations, is easiest solved by the substitution method. You can use the board method to solve mixture problems if there is a total (see Chapter 9), but tried-and-true methods are required if no total is given.

Example 6: How many milliliters of a 26% hydrochloric-acid solution must be mixed with 18 ml of 6% hydrochloric-acid solution to produce a mixture of 18% hydrochloric acid and how many total milliliters are in the final solution?

1. **Read the problem.**
2. **Identify the variables.**

a, the number of milliliters of 26% hydrochloric-acid solution needed

t, the total number of milliliters in the final solution

Because you have two unknowns in this problem, you need two equations.

3. **Translate the equations.**

Reread the problem, translating one sentence at a time into an equation, if possible. The first sentence in the problem translates into both equations.

a. Equation one:

$$a + 18 = t$$

b. Equation two:

$$0.26a + 0.06(18) = 0.18t$$

4. **Solve the system of equations.**

Equation one is easily solved for a. Use the substitution method to solve the system of equations.

$$a + 18 = t$$
$$a + 18 - 18 = t - 18$$
$$a = t - 18$$

Replace a with $(t - 18)$ in equation two.

$$0.26(t - 18) + 0.06(18) = 0.18t$$

$$0.26(t - 18) + 0.06(18) = 0.18t$$
$$0.26t - 4.68 + 1.08 = 0.18t$$
$$0.26t - 3.6 = 0.18t$$
$$0.26t - 0.18t - 3.6 = 0.18t - 0.18t$$
$$0.08t - 3.6 = 0$$
$$0.08t - 3.6 + 3.6 = 0 + 3.6$$
$$0.08t = 3.6$$
$$\frac{0.08t}{0.08} = \frac{3.6}{0.08}$$
$$t = 45$$

Use the equation $a = t - 18$ to solve for the number of milliliters of 26% hydrochloric-acid solution needed.

$$a = 45 - 18$$
$$a = 27$$

Answer the question and include the units: In order to create a 45 milliliter mixture of 18% hydrochloric acid, 27 milliliters of 26% hydrochloric-acid solution must be mixed with 18 milliliters of 6% hydrochloric-acid solution.

5. **Check the solutions.**

   a. Equation one:

$$a + 18 = t$$
$$27 + 18 = 45$$
$$45 = 45 \checkmark$$

   b. Equation two:

$$0.26a + 0.06(18) = 0.18t$$
$$0.26(27) + 0.06(18) = 0.18(45)$$
$$7.02 + 1.08 = 8.1$$
$$8.1 = 8.1 \checkmark$$

Example 7: Kevin has a five-gallon gas can he wishes to fill with a mixture of gas and two-cycle oil for his weed trimmer and other gas-powered lawn tools. Every 50 quarts of gas must be mixed with a quart of two-cycle oil. How much gas and two-cycle oil should Kevin use to fill the five-gallon gas can?

1. **Read the problem.**
2. **Identify the variables.**

   g, the number of quarts of gasoline needed

   t, the number of quarts of two cycle oil needed

   This problem has two unknowns and needs two equations.

3. **Translate the equations.**

   Reread the problem, translating one sentence at a time into an equation, if possible.

   Remember: All measures of volume must use the same units. Convert five gallons into quarts by multiplying five times four because a gallon has four quarts. (In other words, five gallons is the same as 20 quarts.)

a. Equation one:

The first sentence in the problem translates as follows:

$$g + t = 20$$

b. Equation two:

The second sentence translates as follows:

$$g = 50t$$

4. **Solve the system of equations.**

Use the substitution method to solve the system of equations because equation two is already solved for g.

Replace g with $(50t)$ in equation one.

$$50t + t = 20$$
$$51t = 20$$
$$\frac{51t}{51} = \frac{20}{51}$$
$$t = \frac{20}{51}$$

Use the equation $g = 50t$ to solve for the number of quarts of gasoline needed.

$$g = 50\left(\frac{20}{51}\right)$$
$$g = \frac{1,000}{51}$$

Answer the question and include the units: In order to create a 5-gallon mixture of gas and two-cycle oil, $\frac{20}{51}$ quarts of two-cycle oil should be mixed with $19\frac{31}{51}$ quarts of gas.

Note: While the answer with these units is correct, if you actually wanted to buy the gas and two-cycle oil, this answer would not be very practical. Two-cycle oil is sold in 8-ounce containers, so translate quarts to ounces. With 32 ounces in a quart, if you multiply $\frac{20}{51}$ by 32 and write your answer as a decimal, you get a little more than 12.5 ounces. Kevin needs to mix about one and a half 8-ounce bottles of two-cycle oil with enough gas to fill the 5-gallon container.

5. **Check the solutions.**

   a. Equation one:

$$g + t = 20$$
$$\frac{1,000}{51} + \frac{20}{51} = 20$$
$$\frac{1,020}{51} = 20$$
$$20 = 20 \checkmark$$

   b. Equation two:

$$g = 50t$$
$$\frac{1,000}{51} = 50\left(\frac{20}{51}\right)$$
$$\frac{1,000}{51} = \frac{1,000}{51} \checkmark$$

Distance Problems

Infamous train problems, like the one in Example 8, can be worked using the board method when the problem gives a total (see Chapter 9). When the problem does not give a total, use the tried-and-true methods in this section.

Remember: Whether using the board method or tried-and-true methods, using the formula $d = rt$ is the key to success with distance problems.

Example 8: A bullet train in Japan can travel for 2 hours and cover the same distance as a car can cover in 5.4 hours. The bullet train travels 102 miles per hour faster than the car. Find the speed of the bullet train and the car.

1. **Read the problem.**
2. **Identify the variables.**

   b, the speed of the bullet train
   c, the speed of the car

   This problem has two unknowns, so two equations are needed.
3. **Translate equations.**

   Reread the problem, translating one sentence at a time into an equation, if possible. The second sentence is easiest to translate.

a. Equation one:

$$b = c + 102$$

In the first sentence, the bullet train travels the same distance as the car. Therefore, the distances are equal. The distance the bullet train travels can be found by substituting into the formula $d = rt$.

The distance traveled by the bullet train: $d = b(2)$

Similarly, the distance traveled by the car: $d = c(5.4)$

Setting these two distances equal yields the second equation.

b. Equation two:

$$2b = 5.4c$$

4. Solve the system of equations.

Use the substitution method to solve the system of equations; equation one is already solved for b.

Replace b with $(c + 102)$ in equation two.

$$2(c + 102) = 5.4c$$

$$2(c + 102) = 5.4c$$
$$2c + 204 = 5.4c$$
$$2c - 2c + 204 = 5.4c - 2c$$
$$204 = 3.4c$$
$$\frac{204}{3.4} = \frac{3.4c}{3.4}$$
$$60 = c$$
$$c = 60$$

Use the equation $b = c + 102$ to solve for the speed of the bullet train.

$$b = 60 + 102$$
$$b = 162$$

Answer the question and include the units: The speed of the bullet train is 162 miles per hour, and the speed of the car is 60 miles per hour.

5. **Check the solutions.**

   a. Equation one:

$$b = c + 102$$
$$162 = 60 + 102$$
$$162 = 162 \checkmark$$

   b. Equation two:

$$2b = 5.4c$$
$$2(162) = 5.4(60)$$
$$324 = 324 \checkmark$$

Example 9: A boat can travel 225 miles in the same time that a plane can fly 375 miles. The boat travels 60 miles per hour slower than the plane. Find the speed of the boat and the speed of the plane.

1. **Read the problem.**
2. **Identify the variables.**

   b, the speed of the boat
   p, the speed of the plane

   This problem has two unknowns and needs two equations.

3. **Translate the equations.**

   Reread the problem, translating one sentence at a time into an equation, if possible. The second sentence is easiest to translate.

   a. Equation one:

$$b = p - 60$$

   In the first sentence, the boat travels for the same amount of time as the plane. Therefore, the times are equal. Find the time the boat travels by solving the formula $d = rt$ for t. After dividing both sides of the equation by r, the formula for time, t, is: $t = \frac{d}{r}$.

   The time traveled by the boat: $t = \frac{225}{b}$

   Similarly, the time traveled by the plane: $t = \frac{375}{p}$

   Setting these two travel times equal:

$$\frac{225}{b} = \frac{375}{p}$$

   The equation has fractions. The fractions can be cleared by multiplying both sides of the equation by the least common denominator, bp.

$$bp\left(\frac{225}{b}\right) = bp\left(\frac{375}{p}\right)$$

$$\frac{\not{b}p}{1}\left(\frac{225}{\not{b}}\right) = \frac{b\not{p}}{1}\left(\frac{375}{\not{p}}\right)$$

b. Equation two:

$$225p = 375b$$

4. **Solve the system of equations.**

   Use the substitution method to solve the system of equations; equation one is already solved for b.

   Replace b with $(p - 60)$ in equation two.

$$225p = 375(p - 60)$$

$$225p = 375(p - 60)$$

$$225p = 375p - 22{,}500$$

$$225p - 375p = 375p - 22{,}500 - 375p$$

$$-150p = -22{,}500$$

$$\frac{-150p}{-150} = \frac{-22{,}500}{-150}$$

$$p = 150$$

   Use the equation $b = p - 60$ to solve for the speed of the boat.

$$b = 150 - 60$$
$$b = 90$$

   Answer the question and include the units: The speed of the boat is 90 miles per hour, and the speed of the plane is 150 miles per hour.

5. **Check the solutions.**

   a. Equation one:

$$b = p - 60$$
$$90 = 150 - 60$$
$$90 = 90 \checkmark$$

   b. Equation two:

$$225p = 375b$$
$$225(150) = 375(90)$$
$$33{,}750 = 33{,}750 \checkmark$$

Chapter Checkout

Q&A

Solve the following problems. Be sure to answer the question and include the units.

1. Everything at Dollar Haven costs either one dollar or fifty cents. Delia bought 22 items for nineteen dollars (before taxes). How many one-dollar items did she buy? How many fifty-cent items did she buy?
2. Every evening, Maria puts her pocket change into a jar. At the beginning of each year, she takes the jar to the bank to deposit the money in her son's college fund. Last year, she deposited \$101.36. Maria deposited three times as many quarters as nickels, twelve less dimes than nickels, and 26 more pennies than quarters. How many of each type of coin did she deposit?
3. Part of \$3,600 was invested at 8%, and the rest was invested at 7%. The annual simple interest from these investments was \$274. Find the amount invested at each rate.
4. Linda's retirement account is invested in three different funds. Last year, she invested a total of \$240,000 in a fixed account that earned 5%, a mutual fund that earned 6%, and a small-cap fund that earned 8%. She had twice as much invested in the small-cap fund as she did in the mutual fund. She earned a total of \$14,800 in interest. How much did she have invested in each fund?
5. How many ounces of an 8% solution must be mixed with a 18% solution to yield 80 ounces of a 12% solution?
6. Kristi needs to have 32 ounces of a 10.5% carbolic-acid solution. How much 8% carbolic-acid solution and 12% carbolic-acid solution should she mix to get a 10.5% solution?
7. A sports car and a van are traveling the 254 miles between Kansas City and St. Louis on the same highway. The sports car leaves Kansas City at the same time the van leaves St. Louis. The sports car averages 69 miles per hour. The van averages 58 miles per hour. In how many hours will they meet?
8. The Second Lake Pontchartrain Causeway in Louisiana from Mandeville to Metairie is the longest bridge in the world. It is 23 miles long. At 8 a.m., a motorcycle gets on the bridge at Mandeville. At the same time, a SUV gets on the bridge at Metairie. The motorcycle averages 65 mph on the bridge. The SUV averages 50 mph. At what time do they meet?

Answers: 1. Delia bought sixteen one-dollar items and six fifty-cent items. **2.** Maria deposited 330 quarters, 98 dimes, 110 nickels, and 356 pennies. **3.** $2,200 was invested at 8% interest, and $1,400 was invested at 7% interest. **4.** Linda invested $120,000 at 5% interest, $40,000 at 6% interest, and 80,000 at 8% interest. **5.** 48 ounces of the 8% solution must be mixed with 32 ounces of the 18% solution. **6.** Kristi must mix 12 ounces of 8% carbolic-acid solution with 20 ounces of 12% carbolic-acid solution. **7.** The sports car and the van will meet two hours after they began. **8.** The motorcycle and the SUV will meet at 8:12 a.m. (0.2 hours is 12 minutes).

Chapter 11
COMMON ERRORS

Chapter Check-In

- ❑ Variable omission
- ❑ Variable reversal
- ❑ Methods for avoiding common errors

Many methods for avoiding common errors have been sprinkled throughout the book. Those methods are summarized at the end of this chapter. Methods to avoid two other common types of errors — variable omission errors and variable reversal errors — are also discussed in this chapter.

Variable Omission

A **variable omission** error occurs when, in the process of translating a sentence of a word problem into an equation, one of the variables is left out. Example 1 demonstrates this type of error.

Example 1: Out of the sixty English and art classes at Community Junior College, there are five English classes for each art class. How many art classes are taught at Community Junior College?

Note: Without using the board method (covered in Chapters 7 and 9), a common mistake is to translate this problem as follows:

$$60 = 5a. \text{ (This is incorrect.)}$$

This mistake is hard to catch because the equation gives a nice, round answer of 12. The word problem asks only for the number of art classes. If the number of English classes was also asked for, the problem solver would notice that the number of English classes cannot be 60 because the total number of classes is 60.

Here is the correct way to work Example 1:

1. **Read the problem.**
2. **Draw a diagram, if possible.**

   When a problem gives you a total, such as 60 classes, the board method helps avoid a variable omission error.

   There is a total of 60 classes in this problem. Draw a board 60 classes long.

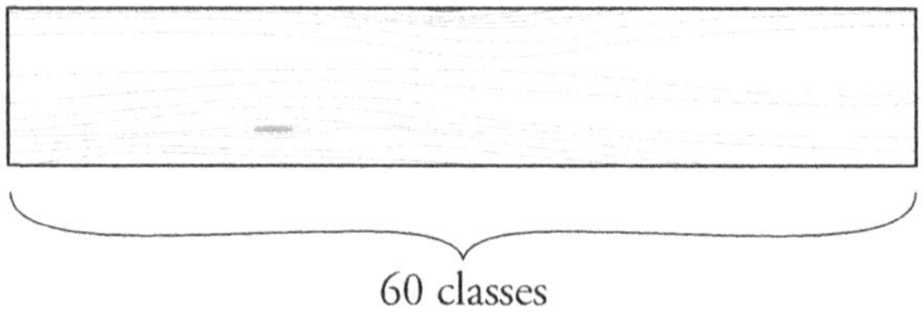

   How many things are added up to get the 60 classes? The number of English classes and the number of art classes are added to get a total of 60. When drawing the board, the tried-and-true method of estimation (see Chapters 7 and 8) is used to place the number of art classes on the smaller portion of the board.

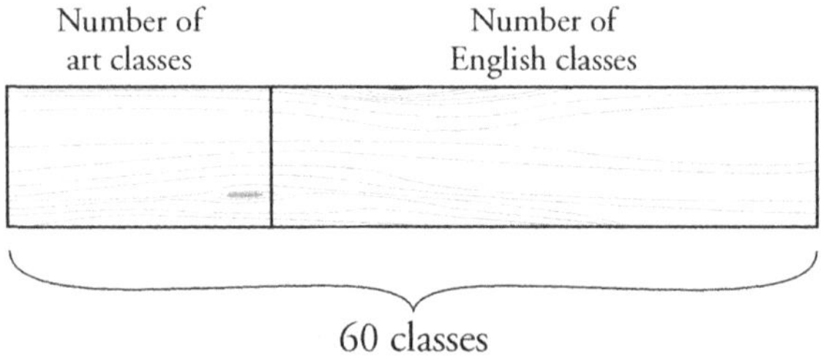

3. **Label the diagram.**

   The text, "there are five English classes for each art class" indicates that the number of English classes is five times the number of art classes. The small piece of the board is labeled *a* for the number of art classes. This method of labeling the board is the same as the tried-and-true method of identifying the variables.

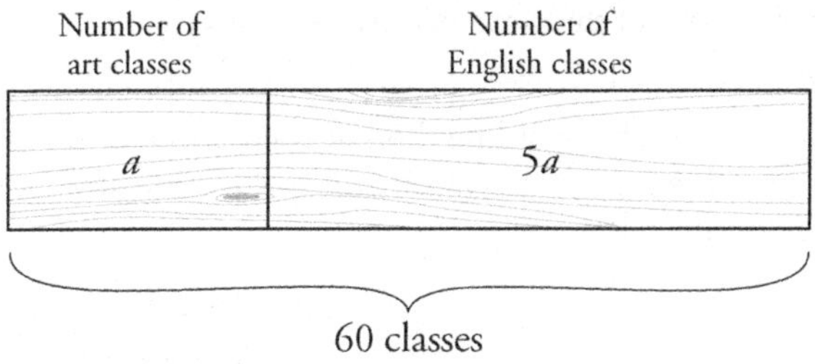

4. **Write the equation.**

$$a + 5a = 60$$

5. **Solve the equation.**

$$6a = 60$$
$$\frac{6a}{6} = \frac{60}{6}$$
$$a = 10$$

Because *a* (for art) was chosen for the unknown variable, the number of art classes is ten. Had the variable *x* been used, you would have to go back to the definition of the variable to be sure which type of classes (art or English) you found.

The number of art classes is ten.

Note: The problem asks only for the number of art classes, not for the number of English classes. A good habit to form is to reread the question after you feel you have the solution. The problem could have easily been worded to ask, "How many English classes are there?"

6. **Check the solution.**

To check the equation, substitute the solution into the equation.

$$a + 5a = 60$$
$$10 + 5(10) = 60$$
$$10 + 50 = 60$$
$$60 = 60 \checkmark$$

Estimation can be used to see whether your solution feels right. Look back at the board and visually determine whether ten art classes is a reasonable answer.

Variable omission error is common in **consecutive integer** problems (discussed in Chapter 7), as shown in Example 2.

Example 2: The sum of three consecutive odd integers is twenty-one. Find the integers.

Note: Without using the board method or tried-and-true methods, some problem solvers skip straight to translating an equation and make the following common mistakes:

$$x + 1 + x + 3 = 21. \text{ (This is incorrect.)}$$

The correct equation is $x + (x + 2) + (x + 4) = 21$

In the incorrect equation, not only is the first consecutive odd integer omitted, but $(x + 1)$ and $(x + 3)$ do not yield odd numbers. If the problem is solved using the board method, the variable omission error is visually obvious when adding three pieces of a board to get a total of 21. The $(x + 1)$ and $(x + 3)$ errors, however, would not be obvious until the solution step. Even if no variable omission occurred, the problem solver could make this mistake and translate the equation as follows:

$$x + x + 1 + x + 3 = 21 \text{ (This is incorrect.)}$$

Solving this equation would yield $5\frac{2}{3}$ as the solution for x, which is not an integer.

This example will be solved using the tried–and-true method, covered in Chapter 8.

1. **Read the problem.**
2. **Identify the variables.**

x, the first odd integer
$x + 2$, the second odd integer
$x + 4$, the third odd integer

3. **Write the equation.**

$$x + (x + 2) + (x + 4) = 21$$

4. **Solve the equation.**

$$3x + 6 = 21$$
$$3x + 6 - 6 = 21 - 6$$
$$3x = 15$$
$$\frac{3x}{3} = \frac{15}{3}$$
$$x = 5$$

Answer the question:

The first odd integer, x, is 5.

The second odd integer, $x + 2$, is 7.

The third odd integer, $x + 4$, is 9.

5. Check the solution.

To check the equation, substitute the solutions into the equation.

$$5 + (5 + 2) + (5 + 4) = 21$$
$$5 + (7) + (5 + 4) = 21$$
$$5 + 7 + (9) = 21$$
$$12 + 9 = 21$$
$$21 = 21 \checkmark$$

Variable Reversal

Another common translating error is called **variable reversal.** When this error is made, the variables are switched with each other.

Example 3: A freshman orientation class has 382 students. The class has six fewer males than females. How many males are in the freshman orientation class?

Note: Many people begin working on this problem by labeling the variable m, the number of male students. This is a reasonable way to begin the problem because the question is asking for the number of males. The mistake is made when translating the equation as:

$$m + (m - 6) = 382 \quad \text{(This is incorrect.)}$$

This gives a solution of 194 male students in the freshman orientation class. This answer sounds reasonable, and many problem solvers would go on to the next problem feeling confident that it was solved correctly. If extra time was taken to solve for the number of female students, the number of female students would be 188. However, the problem said there were fewer males than females. The error made is variable reversal. This error can be avoided by using the board method.

1. Read the problem.

Make helpful markings, as needed.

The class has six fewer males [than] females.

2. Draw a diagram, if possible.

Does the problem have a total? Yes, the total is 382. Draw a board 382 students long.

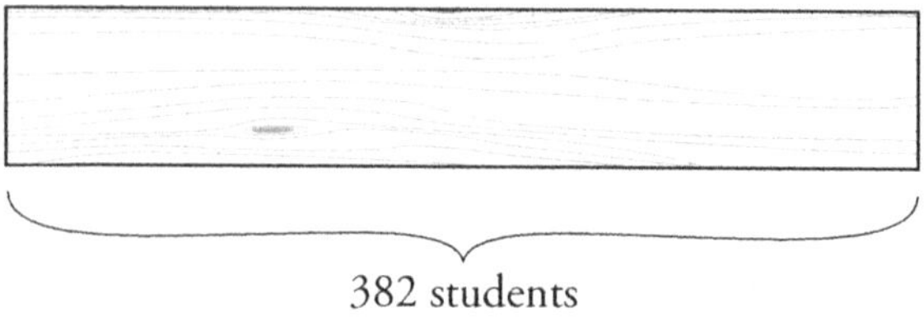

Cut the board into two pieces and place the number of male students on the smaller piece of board because there are fewer males than females.

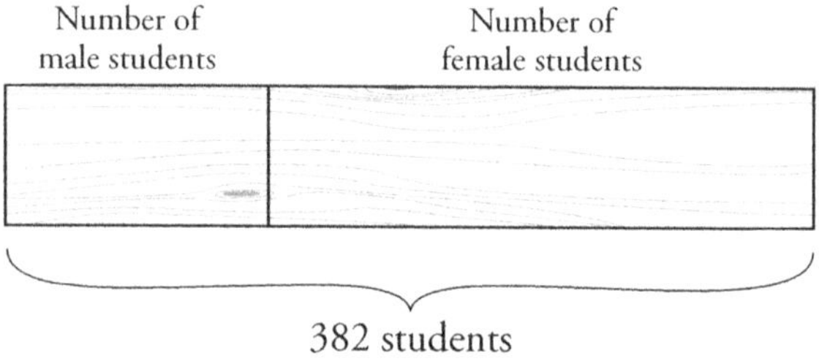

3. Label the diagram.

This can be done in two different, yet correct ways. The problem states the number of male students in terms of the number of female students. The number of female students could be the variable f. The number of male students would then be $f - 6$. The equation would then be solved for f, and that number would then be used to solve for the number of male students because the problem asked only for the number of male students.

Because the problem asks for the number of male students and that number is also represented by the smaller piece of the board, this example will be solved using m for the number of male students.

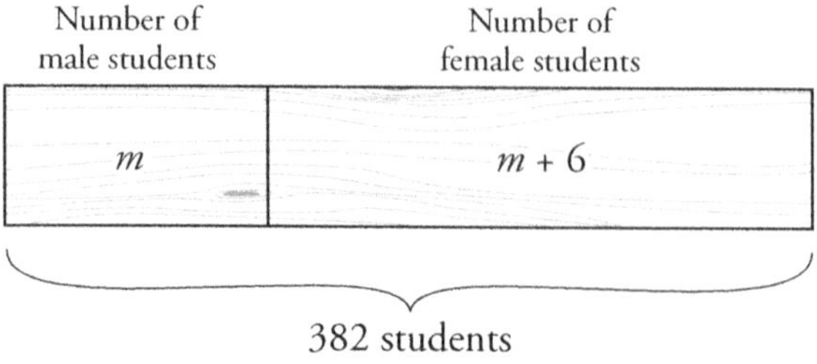

The board gives a visual cue to write the number of female students in terms of m as $m + 6$ even though the problem does not have any keywords that translate to addition.

4. **Write the equation.**

$$m + (m + 6) = 382$$

5. **Solve the equation.**

$$2m + 6 = 382$$
$$2m + 6 - 6 = 382 - 6$$
$$2m = 376$$
$$\frac{2m}{2} = \frac{376}{2}$$
$$m = 188$$

Note: This was the solution for the number of female students when the variable reversal error was made.

The number of male students in the freshman orientation class is 188. It is not necessary to find the number of female students to answer the word problem.

6. **Check the solution.**

To check the equation, substitute the solution into the equation.

$$m + (m + 6) = 382$$
$$188 + (188 + 6) = 382$$
$$188 + 194 = 382$$
$$382 = 382 \checkmark$$

Estimation can be used to see whether your solution feels right. Look back at the board and visually check to see whether 188 male students is a reasonable answer.

Example 4: The Busch Ranch has twelve times as many head of cattle as the Waniska Ranch. If the Busch Ranch has 108 head of cattle, find the number of head of cattle on the Waniska Ranch.

Note: Many people begin working on this problem by labeling the variables as follows:

b, the number of head of cattle on the Busch ranch
w, the number of head of cattle on the Waniska ranch

This is an excellent way to begin the problem. The mistake is made when, using the direct translation method, the first sentence translates as

$$12b = w$$

Substituting 108 for b gives a solution of 1,296 heads of cattle on the Waniska ranch. However, the problem says there are more head of cattle on the Busch ranch than the Waniska ranch. The error made is variable reversal. This error can be avoided by using the tried-and-true method of estimating.

1. **Read the problem.**
2. **Identify the variables.**

   b, the number of heads of cattle on the Busch ranch
   w, the number of heads of cattle on the Waniska ranch

   With two unknowns in this problem, two equations are needed.
3. **Translate equations.**

   Reread the problem, translating one sentence at a time into an equation, if possible. When translating the first sentence, use your estimation skills. Notice there are more cattle at the Busch ranch than the Waniska ranch: b must be more than w.

   The first sentence translates to $b = 12w$

   The second sentence translates to $b = 108$
4. **Solve the system of equations using the substitution method.**

$$\begin{aligned} 108 &= 12w \\ \frac{108}{12} &= \frac{12w}{12} \\ 9 &= w \\ w &= 9 \end{aligned}$$

   Answer the question and include the units with your answer: There are nine head of cattle at the Waniska ranch.
5. **Check the solution.**

$$\begin{aligned} b &= 12w \\ 108 &= 12(9) \\ 108 &= 108 \checkmark \end{aligned}$$

Methods for Avoiding Common Errors

Throughout this book, methods are introduced to improve problem solving skills for math word problems. Common errors and the tools to avoid them are summarized in this section.

A firm foundation

Rather than just simply reading the word problem and writing down a complete equation, the skilled problem solver increases understanding of the problem by following these methods:

- **Identify the variable(s) by looking at the last line of the word problem to find out what is unknown.** Assign the variable a letter of the alphabet that makes sense and write a brief description of the variable. Often, the letter of the alphabet chosen can be the first letter of the unknown value.

  However, using *s* as the letter of the alphabet to represent the variable is ill-advised because an *s* can resemble the number five. In the same way, *t* is also ill-advised because it can resemble a plus sign. If the letter *z* is used, you may want to write Ƶ so that the *z* will not be confused with the number 2.

- **Draw a diagram, whenever possible.** (Chapter 5 discusses how to draw diagrams.) As each side of the diagram is labeled, the translation of the problem is broken into smaller pieces. The equation is then formed by looking at the diagram.

  If a word problem is not easily pictured, ask yourself whether the problem gives you a total. If it has a total, the board method (Chapter 7) breaks the translation of the problem into smaller pieces.

- **Estimate the solution of the word problem.** Take a guess at what you think will be a possible answer. Look at the relationships between variables and get a feel for the solution. Your estimate will be fairly accurate.

Errors in translation

The direct translation strategy works under many circumstances. Often, one sentence can be translated one word at a time into an equation in the exact same order. Errors occur when the order must be changed or parentheses are needed. A change in the order is indicated by the turnaround words:

TO
FROM
THAN

The four hints indicating the need for parentheses, as given in Chapter 2, are as follows:

- A comma in the expression indicates the completion of one operation and the beginning of another. The word IS also acts as a separator, just like a comma. Put parentheses around the completed expression.
- If you see a leading keyword, draw lines under the words before and after the corresponding AND, TO, BY, or FROM. If the underlined portion contains a keyword, put parentheses around the underlined expression.
- When you see two adjacent keywords, separate them by an open parenthesis, and then close the parentheses at the end of the expression.
- If the keyword includes a turnaround word, TO, FROM or THAN, and the expression(s) that is to be turned around has another keyword, that expression(s) must be enclosed in parentheses before turning it around.

Variable omission and variable reversal are also errors in translation; specific methods for avoiding those errors are given in the two preceding sections.

Remember: Some word problems include extraneous information. Not all numbers given in a problem are necessary to solve the problem.

Errors in solving an equation

Avoid errors when solving an equation by showing all steps. A math error is more likely to occur when you solve the problem in your head rather than on paper.

Checking the solution in the original equation can catch errors. Rereading the question with your solution in place of the unknown is another way to check.

If your solution does not pass either one of these checking techniques, go back, find the error, and correct it.

Errors in answering the question

When you have solved the word problem and are ready to go on to the next question, first ask yourself the following questions:

- Did I answer the question?
- Did I include my units?

- Is the answer reasonable?
- Is the answer close to my estimate?

If your solution is drastically different from your estimate, go back and check both your translation and your solution of the equation(s).

Chapter Checkout

Q&A

Solve the following problems. Be sure to answer the question and include units.

1. Out of 56 booths at the State Fair, there are 7 retail sales booths for each food booth. How many food booths are at the State Fair?
2. In the electronics section of a department store, there are 6 DVDs for every video tape. If there is a total of 245 movies, how many video tapes are there?
3. Find two consecutive odd integers such that three times the first added to the second is 46.
4. Find three consecutive odd integers such that the difference between four times the first and the sum of the second and third is zero.
5. XYZ Corporation reserves 546 parking places for employees. There are 54 more parking spots in the garage than on the lot. How many reserved parking spots are in the parking lot?
6. One Monday morning, at checkout lane #5, Mandy noticed four times as many people used plastic than cash or checks. During the three hour shift she worked that morning, she checked out 40 people. How many used plastic?
7. A board measuring 50 ft long is cut into two pieces. The longer piece is four times the size of the smaller piece. How long is the longest piece?

Answers: 1. There are 7 food booths. **2.** There are 35 video tapes. **3.** The two consecutive odd integers are 11 and 13. **4.** The three consecutive odd integers are 3, 5, and 7. **5.** There are 246 reserved parking spots in the parking lot. **6.** Thirty-two people used plastic. **7.** The longest piece is 40 feet long.

CQR REVIEW

Use this CQR Review to practice what you've learned in this book. After you work through the review questions, you're well on your way to achieving your goal of solving math word problems.

Chapter 1

Translate the following English phrases into algebraic expressions.

1. The sum of a number and fourteen
2. A number increased by three
3. The difference between eight and five
4. Seven minus a number
5. The product of a number and six
6. Four times a number
7. The quotient of a number and thirteen
8. Twenty-four divided by two
9. Add nine to a number
10. One subtracted from sixteen
11. Fourteen less than a number
12. Six divided into forty-two
13. The basic turnaround words are ______, ______, and ________.

Chapter 2

Translate the following into algebraic expressions.

14. A number minus four plus six
15. The quotient of a number and two, added to the same number
16. Nine plus a number is subtracted from twenty-eight

17. The difference between a number and three times the number

18. The quotient of a number and thirteen plus the same number

19. The product of six less than a number and four times the same number

20. The sum of five times a number and twice the number

21. Divide a number into a number plus five

22. Three more than twice a number

23. Four times the difference between a number and seven

24. Six less than the total of a number and eleven

Chapter 3

Simplify the following expressions by using the distributive property and combining like terms.

25. $3(x + 4)$

26. $5(x - 6)$

27. $4x - 2x + 7$

28. $9 - 3x + x$

29. $2x + 5(x + 1)$

30. $2(x - 4) - 5x$

31. $9x - 4(x + 2)$

32. $x - 7(x - 2)$

33. $8 - (x + 6)$

34. $3x - (x - 7)$

Chapter 4

Translate and solve the following equations. Be sure to check your solution and translation.

35. Find a number such that three times the number plus two is equal to negative seven.

36. The product of four and a number minus three equals twelve. Find the number.

37. Four times a number is triple the difference between the same number and two. What is the number?

38. When you multiply two by a number subtracted from five, the result is eight. Find the number.

39. Find a number such that two times the number decreased by the sum of the number and one yields six.

Chapter 5

Solve the following word problems. Be sure to answer the questions and include units.

40. Find the area of a triangle with a base of 4 cm and a height of 8 cm.

41. Find the length of one side of a triangle if the perimeter is 15 m, the longer side of the triangle is 8 m, and the shorter side is 3 m.

42. The perimeter of a rectangle is 40 m. The length of the rectangle is 4 m less than three times the width. Find the length and width of the rectangle.

43. The area of a square is 100 m. Find the length of the sides.

44. A farmer has 2,400 ft. of fencing and wants to fence off a rectangular field with the long side bordering a river. He needs no fence along the river. The length is twice as long as the width. Find the dimensions of the field.

45. The length of the side of a square is equal to three times the length of the side of an equilateral triangle. If the perimeter of the square is 48 in., find the length of the side of the equilateral triangle.

46. A board measuring 50 in. long is cut into two pieces. The longer piece is four times the size of the smaller piece. How long is the longer piece?

Chapter 6

Solve the following problems for the unknown. Be sure to answer the questions and include units on the word problems.

47. $\frac{8}{9} = \frac{x}{45}$

48. $\frac{16}{y} = \frac{2}{3}$

49. Angela has been making 7 out of every 8 free throws she attempts. If she can keep up this same rate in a free-throw competition, how many free throws can she expect to make if she is given 40 chances?

50. At the end of a regular week, John has noticed that he has twice as many loads of permanent-press clothes as jeans to be washed. If he has two loads of jeans this week, how many loads of permanent-press clothes should he expect to wash?

51. 14 is 50% of what number?

52. 135 is what percent of 300?

53. Melvin's is having a 25% off sale. How much will a $36 sweater be marked down?

54. Richele's bill at Casa Bonita restaurant was $12.50. The waitress did an excellent job, and Richele wants to tip her 20%. How much money should Richele leave for the tip?

Chapter 7

Solve the following word problems. Be sure to answer the questions and include units.

55. A board measuring 56 in. was cut into three pieces. The longest piece is two times the middle-size piece. The shortest piece is ½ the size of the middle-size piece. How long is the middle-size piece?

56. A piece of string was cut into two pieces. The longer piece was three less than three times the shorter. If the string was 13 inches long, how long is the longer piece?

57. Find three consecutive integers such that the sum of the integers equals the difference between four times the third integer and one.

58. Find two consecutive even integers such that the sum of the first even integer divided by six and nine is equal to the second even integer divided by two.

59. Find two consecutive odd integers such that the second odd integer added to the product of three and the first integer equals thirty.

60. A poll was taken of 540 registered voters. They were asked whether they favored the president's foreign policy. Some of the voters were undecided. Twice as many voters polled voted no as those who voted undecided. Three times as many voters favored the president's foreign policy as were undecided. How many favored the president's foreign policy?

61. Chocolate Bonbon Co. has been making bonbons for Valentine's Day for years. The company purchased a new bonbon machine this year because it is concerned that the old one will break soon. The old machine can fill an order of 1,000 boxes in 7 hours. The new machine can fill an order of 1,000 boxes in 5 hours. How many hours will it take to fill an order of 1,000 boxes if both machines are running?

Chapter 8

Solve the following word problems. Be sure to answer the questions and include units.

62. Find two consecutive integers such that if the quotient of the first integer and three is added to seven, the result is equal to the second integer.

63. Find two consecutive even integers such that the second even integer minus two times the first even integer is negative six.

64. Find two consecutive odd integers such that the second odd integer minus two times the first odd integer is seven.

65. The average body temperature of humans is 37 degrees Celsius. The average person is uncomfortable in a room that is 37 degrees Celsius. Most people feel comfortable in a room that is from 16 to 11 degrees Celsius cooler than their body temperature Find the range of temperatures in which most people are comfortable.

66. It is said that a healthy person should drink 8 cups of water a day. That is one half of a gallon of water per day. A healthy horse should have more than twenty times as much water. How many gallons of water should a healthy horse have?

67. Cedric can weld 200 flanges in 18 hours. It takes Mike 20 hours to weld 200 flanges. If they work together, how long will it take them to weld 200 flanges?

Chapter 9

Solve the following word problems. Be sure to answer the questions and include units.

68. Ashley has $6.30 in a change jar. She saves only dimes, nickels, and pennies. She has twice as many dimes as nickels and five times as many pennies as nickels. How many of each coin does she have?

69. Ray invested $23,400 in stocks and commodities. In one year's time, he earned $1,638. He averaged 5% simple interest with the stocks and 8% simple interest with the commodities. How much did he invest in commodities?

70. In chemistry class, Joshua was asked to make 40 ounces of a 16% solution of phosphoric acid. He was given a 10% solution of phosphoric acid and a 20% solution of phosphoric acid. How many ounces of each solution should he mix to get the correct solution?

71. Two trains leave the station at the same time going in opposite directions. After 2 hours, the trains are 250 miles apart. One train averages 5 mph faster than the other. Find the speed of each train.

72. A train leaves the station at 10 a.m. traveling north. Another train leaves the station at 10 a.m. traveling south. At 2 p.m., they are 456 miles apart. The train traveling north averaged 10 mph slower than the train traveling south. Find the speed of each train.

Chapter 10

Solve the following word problems. Be sure to answer the questions and include units.

73. Jasmine's Junior Achievement project was to create Jasmine's Jewelry, a business that makes necklaces and bracelets. The necklaces sell for $6 each, and the bracelets sell for $4. In one week of sales, she made $114. She sold a total of 21 items. How many of each did she sell?

74. Regular tickets to hear Las Pappas Fritas cost $10, and seats on the floor cost $15. The concert hall holds 540 people, and the concert is sold out. The take from the tickets totaled $6,550. How many of each type of ticket was sold?

75. Avery invested $3,100. He diversified his investment by investing in a fixed interest account and in small cap funds. Over the past year, the fixed interest account yielded 4.5% simple interest, while the small cap funds averaged 7.2% simple interest. He made $196.20 in interest last year. How much did he have invested in the fixed interest account?

76. Twenty-four-karat gold is pure gold. Eighteen-karat gold is 18 parts pure gold and 6 parts other metal. In other words, 18-karat gold is 75% gold. Twelve-karat gold is 12 parts gold and 12 parts other metals, or 50% gold. A jeweler wants to make 18-karat gold jewelry from old and damaged 24- and 12-karat gold jewelry. How many ounces of each should the jeweler mix to yield 24 ounces of 18-karat gold?

77. The distance between El Paso and Beaumont is 840 miles. An RV leaves Beaumont and travels west on I-10. At the same time, a car leaves El Paso and travels east on I-10. Even with multiple stops for gas, the RV averages 44 miles per hour. The car averages 61 miles per hour. In how many hours will they meet?

Chapter 11

Solve the following word problems. Be sure to answer the questions and include units.

78. Austin has 24 rabbits in his rabbitry. There are three times as many does as bucks. How many of his rabbits are does?

79. Amelia has an extensive collection of 2,400 books. The number of paperback books is four times the number of hardback books. How many paperback books does she have?

80. There are 42 players on a high school football team. One third of the players weigh less that 200 pounds. How many weigh 200 pounds or more?

81. Find three consecutive odd integers such that the sum of the integers is negative three times the first odd integer.

82. A board measuring 18 ft. long is cut into two pieces. The longer piece is two times the length of the shorter piece. Find the length of the longer piece.

Answers: **1.** $x + 14$ **2.** $x + 6$ **3.** $8 - 5$ **4.** $7 - x$ **5.** $n \times 6$ **6.** $4x$ **7.** $x \div 13$ **8.** $24 \div 2$ **9.** $x + 9$ **10.** $16 - 1$ **11.** $x - 14$ **12.** $42 \div 6$ **13.** TO, FROM, and THAN **14.** $x - 4 + 6$ **15.** $x + (x \div 2)$ **16.** $28 - (9 + x)$ **17.** $x - (3x)$ **18.** $x \div (13 + x)$ **19.** $(x - 6) \times (4x)$ **20.** $(5x) + (2x)$ **21.** $(x + 5) \div x$ **22.** $(2x) + 3$ **23.** $4(x - 7)$ **24.** $(x + 11) - 6$ **25.** $3x + 12$ **26.** $5x - 30$ **27.** $2x + 7$

28. $9 - 2x$ **29.** $7x + 5$ **30.** $-3x - 8$ **31.** $5x - 8$ **32.** $-6x + 14$ **33.** $-x + 2$ **34.** $2x + 7$ **35.** $3x + 2 = -7; x = -3$ **36.** $4(x - 3) = 12; x = 6$ **37.** $4x = 3(x - 2); x = -6$ **38.** $2(5 - x) = 8; x = 1$ **39.** $2x - (x + 1) = 6; x = 7$ **40.** The area of the triangle is 16 cm^2. **41.** The length of the other side of the triangle is 4 meters. **42.** The width of the rectangle is 6 meters; the length of the rectangle is 14 meters. **43.** The length of side of the square is 10 meters. **44.** The width of the field is 600 feet; the length of the field is 1,200 feet. **45.** The length of the side of the equilateral triangle is 4 inches. **46.** The longer piece is 40 inches long. **47.** $x = 40$ **48.** $y = 24$ **49.** Angela can expect to make 35 of the 40 free throws. **50.** John should expect to wash 4 loads of permanent press clothes. **51.** 14 is 50% of 28. **52.** 135 is 45% of 300. **53.** The $36 sweater will be marked down $9. **54.** Richele should leave a $2.50 tip. **55.** The middle-size piece is 16 inches. **56.** The longer piece is 9 inches long. **57.** The three consecutive integers are –4, –3, and –2. **58.** The two consecutive even integers are 24 and 26. **59.** The two consecutive odd integers are 7 and 9. **60.** 270 of the voters polled favored the president's foreign policy. **61.** It will take $2\frac{11}{12}$ hours or 2 hours and 50 minutes to complete the order for 1,000 boxes if both machines are running. **62.** The first integer is 9, and the second integer is 10. **63.** The first even integer is 8, and the second even integer is 10. **64.** The first odd integer is –5, and the second odd integer is –3. **65.** Most people are comfortable in a room that is between 21 and 26 degrees Celsius. **66.** A healthy horse should drink at least 10 gallons of water. **67.** If Cedric and Mike work together, it will take approximately 9.47 (or $9\frac{1}{2}$) hours to weld the 200 flanges. **68.** Ashley has 105 pennies, 21 nickels, and 42 dimes. **69.** Ray invested $15,600 in commodities. **70.** Joshua should mix 24 ounces of the 20% solution with 16 ounces of the 10% solution. **71.** The faster train averages 65 mph, and the slower train averages 60 mph. **72.** The slower train is traveling north at 52 mph; the faster train is traveling south at 62 mph. **73.** Jasmine's Jewelry sold 15 necklaces and 6 bracelets. **74.** 310 regular tickets were sold at $10 each, and 230 tickets for floor seating were sold at $15 each. **75.** Avery invested $1,000 in the fixed interest account. **76.** The jeweler should mix 12 ounces of 24-karat gold with 12 ounces of 12-karat gold to get 24 ounces of 18-karat gold. **77.** The car and the RV will meet in 8 hours. **78.** Austin has 18 does in his rabbitry. **79.** Amelia has 1,920 paperback books in her collection. **80.** 28 players on the high school football team weigh 200 pounds or more. **81.** The three consecutive odd integers are –1, 1, and 3. **82.** The longest piece is 12 feet long.

CQR RESOURCE CENTER

CQR Resource Center offers the best resources available in print and online to help you study and review the core concepts of math word problems. The resourses in this section are listed in order of author preference. You can find additional resources, plus study tips and tools to help test your knowledge, at `www.cliffsnotes.com`.

Books

This CliffsQuickReview book is one of many great books about math word problems. If you want some additional resources, check out these other publications:

Algebra For Dummies, by Mary Jane Sterling, explains algebra concepts in an easy way, covering fractions, exponents, factoring, equations, graphing, and word problems. Wiley. $19.99.

CliffsQuickReview Algebra I, by Jerry Bobrow, reviews monomials, inequalities, functions, roots, and radicals, and has a whole chapter devoted to word problems. Wiley. $9.99.

CliffsQuickReview Basic Math and Pre-Algebra, by Jerry Bobrow, reviews whole numbers, fractions, decimals, and percents integers, plus has a whole chapter devoted to word problems. Wiley. $9.99.

Prealgebra: A Worktext, by D. Franklin Wright, guides the adult reader through basic pre-algebra topics. This college textbook does an excellent job of integrating word problems throughout the book. Hawkes Publishing. $75.40.

Contemporary's Number Power: Real World Approach to Math: Word Problems, by Kenneth Tamarkin, gives you examples and a number of word problems with solutions. This is book number six in a series of seven books. NTC/Contemporary Publishing Co. $11.50.

Wiley also has three Web sites that you can visit to read about all the books we publish:

- `www.cliffsnotes.com`
- `www.dummies.com`
- `www.wiley.com`

Internet

Visit the following Web sites for more information about math word problems:

Ms. Linquist: The Free Algebra Tutor for Word Problems—www.algebratutor.org—is a site funded by the National Science Foundation and the Spencer Foundation to teach students how to translate word problems into algebraic equations.

Algebra Homework Help—www.algebra.com—has interactive solutions for word problems and other support for algebraic concepts.

Translating Word Problems—www.purplemath.com/modules/translat.htm—is a tutorial designed to help you translate word problems into the correct equations.

Understanding Algebra—www.jamesbrennan.org/algebra—offers an online book written by James Brennan; one chapter is devoted to word problems.

Webmath—www.webmath.com—offers links to many topics in math. The algebra link leads to specific sites that help you with word problems.

Math Story Problems—www.mathcats.com/storyproblems.html—offers an assortment of story problems with different levels of difficulty, ranging from kitten(easiest) to tiger(hardest).

Next time you're on the Internet, don't forget to drop by `www.cliffsnotes.com`. We've created an online Resource Center that you can use today, tomorrow, and beyond.

GLOSSARY

abstract Intangible, not visible or touchable.

acute angle An angle that measures less than 90 degrees.

acute triangle A triangle in which all three angles are acute angles; in other words, each angle measures less than 90 degrees.

addition keywords Words that indicate addition.

addition method Sometimes called the elimination method, it is a method for solving a system of two equations. One or both of the equations needs to be modified so that when the two equations are added, one of the variables is eliminated.

addition property of equations An equation is still true if the same term is added to (or subtracted from) both sides of an equation.

adjacent keywords Two keywords that are next to each other (even if a word such as "the" appears between the keywords).

algebraic equation A statement that two algebraic expressions are equal.

algebraic expression A collection of numbers, variables, operations, and grouping symbols.

amount The number that is a percentage of the base. It is the number in the numerator in a percent ratio:

$a/b = p/100$ (see also definitions of base and percent)

analogies Similarities in relationships between two items.

angle Two rays sharing the same endpoint. (A ray is a part of a line that looks like an arrow. It begins at one point and goes on forever in the opposite direction.)

area The amount of surface enclosed by a closed figure.

base The number that is in the denominator in a percent ratio:

$a/b = p/100$ (see also definitions of amount and percent)

base (of a triangle) One of the three sides of a triangle that is perpendicular to the height.

board method This method can be used to help translate a word problem into an equation if there is a total given in the problem.

clearing fractions A method of simplifying an equation by multiplying both sides of the equation by the least common denominator before solving it. This method results in an equation with only whole numbers and no fractions.

coefficient The number in front of a variable. For example, 4 is the coefficient in the term $4x$.

combining like terms The process of adding or subtracting like terms.

commutative property of addition The order of the terms does not change the sum.

complement The percentage that would add to 100%. For example, 42% is the complement of 58%.

concrete Tangible, easily understood.

consecutive even integers Integers that are adjacent in an ordered list of even integers. For example, 4 and 6 are two consecutive even integers.

consecutive integers Integers that are adjacent on a number line. For example, 7, 8, and 9 are three consecutive integers.

consecutive odd integers Integers that are adjacent in an ordered list of odd integers. For example, 1 and 3 are two consecutive odd integers.

direct translation strategy This strategy is used when you can translate each word, one at a time in the same order as written, into its corresponding algebraic symbol.

distance problems Word problems that involve traveling a distance. The formula $d = rt$ is used for these problems.

distributive property of multiplication over addition The property that allows a number or term to be distributed (by multiplication) to the sum of two terms in parentheses, for example, $a(b + c) = ab + ac$.

division keywords Words that indicate division.

draw a picture Create a drawing that helps visualize the word problem.

elimination method Sometimes called the addition method, it is a method for solving a system of two equations. One or both of the equations needs to be modified so that when the two equations are added, one of the variables is eliminated.

equation Two expressions set equal to each other. The easiest way to differentiate between an expression and an equation is that the equation has an equal sign.

equilateral triangle A triangle with all three sides of equal length.

estimate An educated guess of the solution. This guess is made before setting up an equation and solving the problem.

evaluation To follow the order of operations and determine the value.

even integers Integers that are even numbers.

even number A number divisible by two.

expression A collection of constants, variables, symbols of operations, and grouping symbols.

grouping symbols Usually parentheses, but grouping is also indicated by brackets and braces.

harmless parentheses Parentheses that are not necessary but that do not do any harm in the expression.

height (of a triangle) A line segment that is perpendicular to the base of a triangle, with one endpoint being the opposite vertex.

homogeneous units When both numbers are measured with the same units.

identify the variable(s) The process of choosing a letter to represent the unknown value and describing that variable.

identity property of multiplication Any number times one has the same value.

implied multiplication Multiplication is implied when a number is placed next to a variable, or when a number is placed next to an expression surrounded by parentheses.

integer A counting number or a negative whole number.

investment problems Word problems that involve investing money at a simple interest rate.

isosceles triangle A triangle with two sides of equal length.

keywords Words that indicate a mathematical operation.

leading keywords If the first word in an English phrase indicates an operation, it is a leading keyword; it "leads" the expression.

least common denominator The least common multiple of all the denominators in the problem.

least common multiple The smallest of the common multiples of more than one number. For example, 12 and 24 are both multiples of 3 and 4. 12 is the least common multiple of 3 and 4.

like terms Terms with the same variables raised to the same powers.

linear equation An equation that can be put in the form $Ax + By + C = 0$.

mixture problems Word problems that involve mixing two or more items to create a new mixture. Often, these are chemistry problems, but they can include problems in which two different types of nuts are mixed to create mixed nuts, and so on.

money problems Word problems that concern money.

multiple equation word problems Word problems that can be translated into more than one equation.

multiple operations An expression has multiple operations when you see more than one symbol for addition, subtraction, multiplication, and/or division.

multiples of a number An infinite list of the products of a number and each whole number. For example, the multiples of 3 are 3, 6, 9, 12, 15, 18 . . .

multiplication keywords Words that indicate multiplication.

multiplication property of equations An equation is still true if both sides of the equation are multiplied by (or divided by) the same term.

obtuse angle An angle that measures more than 90 degrees and less than 180 degrees.

obtuse triangle A triangle with one obtuse angle.

odd A number that is not divisible by two.

odd integers Integers that are odd numbers.

order of operations When an expression has multiple operations, they must be performed in the following order: 1) operations within parentheses; 2) exponents; 3) multiplication and division, from left to right; and 4) addition and subtraction, from left to right.

percent(s) The part per one hundred in the percent proportion:

$a/b = p/100$ (see also definitions of base and percent)

percentages A given part in every hundred.

perimeter The sum of the lengths of the sides of any closed figure.

perpendicular Two line segments are perpendicular if they meet at a right angle, 90 degrees.

Polya's four-step process A process of problem solving published in 1945 by George Polya. The steps are as follows: 1) Understand the problem; 2) Devise a plan; 3) Carry out the plan; 4) Look back over the results.

proportion(s) Two ratios set equal to each other.

ratio A comparison of two quantities by division.

rectangle A four-sided closed figure in which the opposite sides are parallel and of equal length. All pairs of adjacent sides of a rectangle meet at right angles.

right angle An angle that measures 90 degrees.

right triangle A triangle with one right angle.

scalene triangle A triangle with all three sides of different lengths.

simplified An algebraic expression is simplified by using the distributive property and combining like terms.

square A square is a special type of rectangle in which all sides are equal in length.

substitution method A method for solving a system of two equations. One of the equations needs to be solved for one of the variables. That expression is then substituted into the other equation for the variable. The resulting equation has only one unknown and can be solved by methods taught in Chapter 4.

subtraction keywords Words that indicate subtraction.

systems of equations More than one equation that are solved simultaneously.

term The addends of an expression.

translate Change a phrase written in English to an algebraic expression, using the correct symbols.

triangle A three-sided closed figure.

tried-and-true method A method for solving word problems that has been used for many years. Some examples are Polya's four-step process, identifying variables, and estimation.

turnaround words Words that indicate a change in order from left to right. The expressions are turned around, and the second English phrase becomes the first algebraic expression (and visa versa). The basic turnaround words are, TO (including INTO), FROM, and THAN.

units The method of measurement used for the numbers in a word problem. For example: feet, inches, dollars, degrees, and so on.

variable A symbol used to represent an unknown number, often x or n.

variable omission An error that occurs when, in the process of translating a sentence in the word problem into an equation, one of the variables is left out.

variable reversal This error is made if the variables are switched with each other.

vertices The three endpoints of the line segments that make up a triangle.

work problems Word problems that involve two people or machines working together at different rates to complete one whole job.

Index

D

E

L

M

N

O

P

U

V

W

Y

Notes

Notes

Made in the USA
Monee, IL
04 November 2026

67384743R00125